Rheology for Chemists

An Introduction

Preface

Every day we are all concerned with the rheological response of a variety of materials because we have to pour, spread or reshape them. Hopefully it will be in a peaceful situation such as standing on the seashore with water lapping around our feet as they gently sink into the sand, and not in the situation where flows are produced by erupting volcanoes. Rheological behaviour is fundamental to our existence, demonstrated by the way in which our blood flows and by our very conception, where the way in which the flow properties of polyelectrolyte gels change due to changes in pH is a critical factor.

In the workplace, many chemists have the problem of formulating materials into a convenient form. Although we could use as an example foods, pharmaceuticals or cleaning materials, let us consider a decorative paint which serves to illustrate the range of responses that we demand. The function of a paint is two-fold. We wish to apply a polymer film to protect the surface that we are painting and secondly, the paint acts as a carrier for pigments to give a decorative finish. In the can we want the pigment particles to remain suspended and to this end we produce a weak gel. A by-product of this is the non-drip behaviour. On application the paint must easily 'thin' to be readily laid on, then we have the problem of levelling and runs prior to drying. The former is driven by surface tension forces whilst the latter is the response to gravitational forces on the film. Hence we require the gelation to start to reoccur but at a rate at which sufficient levelling will take place. So what exactly are we asking this material to do? Firstly it should be a soft solid that melts when we apply a force with a brush or roller, and then it must resolidify at a controlled rate. How do we achieve this? Well, not by magic, but by chemistry. We control the interactions between the molecules and the particles in the paint so that the best structure and diffusional timescales for our purposes are produced. It is the purpose of this book to clarify this process. Not just with paint, of course, but with any formulation.

The excitement in the study of rheology is in seeing how the timescales are so important in how our materials behave. For the chemist it is rewarding to see how the controlling factor is the same intermolecular forces that we have been trained to manipulate. Now we will have to work in terms of stresses and strains and use some simple algebra in order to enable us to describe or predict behaviour. As this is an introductory text, derivations will only be given where they are straightforward and provide greater understanding. For more complex results, the important relationship is given and the enthusiast can find more detail from the appropriate references. The algebra is a simple compact shorthand notation that enables us to summarise the behaviour; much more important is the understanding of the mechanisms involved as it is these that give us the 'feel' for a subject. It is this that we wish to promote and to this end we restrict ourselves to the simple experiments that we would normally carry out in the laboratory and do not tackle the complex flows that may be important for engineering applications.

The format of the book is very straightforward. The subject with its essential terminology is introduced in the first chapter. The following two chapters develop the ideas for the limiting behaviour, *i.e.* when we are not too concerned with the timescales. The next two chapters develop these ideas further as the temporal behaviour comes to the fore. Finally we move into non-linear behaviour. Most readers will feel at home here as we discuss the types of experiment that they are carrying out every day. Our aim is that every chapter should be as self-contained as possible and so we revisit basic ideas and extend them where necessary, with the intention that the depth of understanding will increase as the reader progresses through the book. Above all our interest is in how atoms and molecules interact to control the handling properties of materials. Many of the systems of importance to the chemist are polymeric and particulate systems and discussion of these takes up the lion's share of the book, but it is the same forces that occur between simple molecules that we must consider in these cases too. Few undergraduate or graduate programmes have much, if any, discussion of rheology, polymers and colloids and we see this volume as the starting point for repairing this omission.

Jim Goodwin, Roy Hughes
November 1999

Contents

CHAPTER 1

Introduction

The study of rheology is the study of the deformation of matter resulting from the application of a force. The type of deformation depends on the state of matter. For example, gases and liquids will flow when a force is applied, whilst solids will deform by a fixed amount and we expect them to regain their shape when the force is removed. In other words we are studying the '*handling properties* of *materials*'. This immediately reminds us that we must consider solutions and dispersions and not simply pure materials. In fact, the utility of many of the materials we make use of every day is due to their rheological behaviour and many chemists are formulating materials to have a particular range of textures, flow properties, etc. or are endeavouring to control transport properties in a manufacturing plant. Interest in the textures of materials such as a chocolate mousse or a shower gel may be of professional interest to the chemist in addition to natural curiosity. How do we describe their textures quantitatively? What measurements should we make? What is the chemistry underlying the texture so that we may control it? All these questions make us focus on rheology.

The aim of this text is to enable the reader to gain an understanding of the physical origins of viscosity, elasticity and viscoelasticity. The route that we shall follow is to introduce the key concepts through physical ideas and analogues that are familiar to chemists and biologists. Ideas from chemical kinetics, and infrared and microwave spectroscopy are invariably covered in some depth in many science courses and so should aid the understanding of rheological processes. The mathematical content is kept to the minimum necessary to give us a quantitative description of a process, and we have taken care to make any manipulations as transparent as possible.

There are two important underlying ideas that we shall return to throughout this work. Firstly, we should be aware that intermolecular forces control the way in which materials behave. This is where the chemical nature is controlling the physical response. The second is the importance of the timescale of our observations, and here we may become aware of different physical responses if our experiments are carried out at different times. The link between the two arises through the *structure* that is the consequence of the forces and the timescale for changes by microstructural motion resulting from thermal or mechanical energy. What is so exciting about rheology is the insights that we can gain into the origins of the behaviour of a wide variety of systems in our everyday mechanical world.

1.1 DEFINITIONS

1.1.1 Stress and Strain

The *stress* is simply defined as the force divided by the area over which it is applied. Pressure is a *compressive bulk stress*. When we hang a weight on a wire, we are applying an *extensional stress* and when we slide a piece of paper over a gummed surface to reach the correct position, we are applying a *shear stress*. We will focus more strongly on this latter stress because most of our instruments are designed around this format. The units of stress are Pascals.

When a stress is applied to a material, a deformation will occur. In order to make calculations tractable, we define the *strain* as the relative deformation, *i.e.* the deformation per unit length. The length that we use is the one over which the deformation occurs. This is illustrated in Figures 1.1 and 1.2.

There are several features of note in Figures 1.1 and 1.2:

1. The elastic modulus is constant at small stresses and strains. This linearity gives us Hooke's Law[1], which states that the stress is directly proportional to the strain.

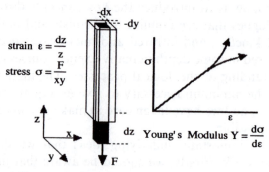

$$\text{strain } \varepsilon = \frac{dz}{z}$$

$$\text{stress } \sigma = \frac{F}{xy}$$

$$\text{Young's Modulus } Y = \frac{d\sigma}{d\varepsilon}$$

Figure 1.1 *Extensional strain at constant volume.* $\varepsilon = \gamma_{zz} = (\gamma_{xx} + \gamma_{yy})$

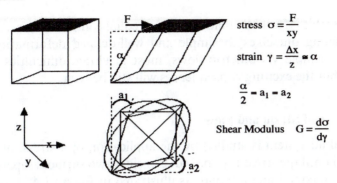

$$\text{stress } \sigma = \frac{F}{xy}$$

$$\text{strain } \gamma = \frac{\Delta x}{z} \approx \alpha$$

$$\frac{\alpha}{2} = a_1 = a_2$$

$$\text{Shear Modulus } \quad G = \frac{d\sigma}{d\gamma}$$

Figure 1.2 *Shear strain* $\gamma = \gamma_{xz} = \gamma_{zx}$

2. At high stresses and strains, non-linearity is observed. Strain hardening (an increasing modulus with increasing strain up to fracture) is normally observed with polymeric networks. Strain softening is observed with some metals and colloids until yield is observed.

3. We should recognise that stress and strain are tensor quantities and not scalars. This will not present any difficulties in this text but we should bear it in mind because the consequences can be both dramatic and useful. To illustrate the mathematical problem, we can think about what happens when we apply a strain to an element of our material. The strain is made up of three orthogonal components which can be further subdivided into three elements, each of which is lined up with one of our axes. This is shown in Figure 1.3.

Figures 1.2 and 1.3 show how, if we apply a simple shear strain, γ, in our

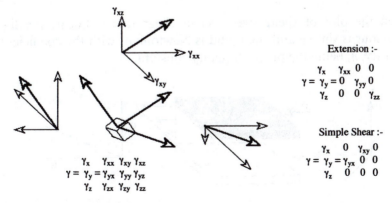

Extension :-

$$\gamma = \begin{matrix} \gamma_x \\ \gamma_y \\ \gamma_z \end{matrix} = \begin{matrix} \gamma_{xx} & 0 & 0 \\ 0 & \gamma_{yy} & 0 \\ 0 & 0 & \gamma_{zz} \end{matrix}$$

Simple Shear :-

$$\gamma = \begin{matrix} \gamma_x \\ \gamma_y \\ \gamma_z \end{matrix} = \begin{matrix} \gamma_{yx} & 0 & \gamma_{xy} & 0 \\ & 0 & 0 \\ & 0 & 0 & 0 \end{matrix}$$

$$\gamma = \begin{matrix} \gamma_x \\ \gamma_y \\ \gamma_z \end{matrix} = \begin{matrix} \gamma_{xx} & \gamma_{xy} & \gamma_{xz} \\ \gamma_{yx} & \gamma_{yy} & \gamma_{yz} \\ \gamma_{zx} & \gamma_{zy} & \gamma_{zz} \end{matrix}$$

Figure 1.3 *Strain and stress are tensors*

rheometer this is formally made up of two equal components, γ_{xy} and γ_{yx}. By restricting ourselves to simple and well-defined deformations and flows, *i.e.* simple viscometric flows, most algebraic difficulties will be avoided but the exciting consequences will still be seen.

1.1.2 Rate of Strain and Flow

When a fluid system is studied by the application of a stress, motion is produced until the stress is removed. Consider two surfaces separated by a small gap containing a liquid, as illustrated in Figure 1.4. A constant shear stress must be maintained on the upper surface for it to move at a constant velocity, u. If we can assume that there is no slip between the surface and the liquid, there is a continuous change in velocity across the *small* gap to zero at the lower surface. Now in each second the displacement produced is x and the strain is

$$\gamma = \frac{x}{z} \tag{1.1}$$

and as $u = \frac{dx}{dt}$, we can write the *rate of strain* as

$$\frac{d\gamma}{dt} = \frac{u}{z} \tag{1.2}$$

The terms *rate of strain*, *velocity gradient* and *shear rate* are all used synonymously and Newton's dot is normally used to indicate the differential operator with respect to time. For large gaps the rate of strain will vary across the gap and so we should write

$$\dot{\gamma} = \frac{du}{dz} \tag{1.3}$$

When the plot of shear stress versus shear rate is linear, the liquid behaviour is simple and the liquid is Newtonian[2] with the coefficient of viscosity, η, being the proportionality constant.

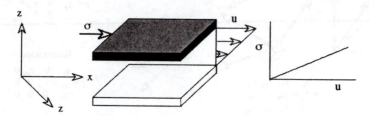

Figure 1.4 *A velocity gradient produced when a fluid is sheared*

When a flow is used which causes an extension of a liquid, the resistance to this motion arises from the *extensional viscosity*, η_ε, and the extension rate is $\dot{\varepsilon}$. Extensional flows require an acceleration of the fluid as it thins and so *steady flows are never achieved*. This means that microstructural timescale is particularly important. Many practical applications involve extensional flows, frequently with a shear component. For example spraying, spreading and roller coating are common ways of applying products from the food, pharmaceutical, paint and printing industries. Although the analysis may be carried out as though the materials are continua with uniform properties, the control comes from an understanding of the role of molecular architecture and forces.

1.2   SIMPLE CONSTITUTIVE EQUATIONS

1.2.1   Linear and Non-linear Behaviour

It is easy to write down an algebraic relationship that describes the simpler forms of rheological behaviour. For example

for a Hookean solid $$\sigma = G\gamma \tag{1.4}$$

and a Newtonian liquid $$\sigma = \eta\dot{\gamma} \tag{1.5}$$

These equations should fully describe the stress–strain–time relationship for the materials over the full range of response. However, the range over which such *linear* behaviour is observed is invariably limited. Usually large stresses and strains or short times cause deviations from Equations 1.4 or 1.5.

As the behaviour becomes more complicated, more parameters are required to fit the experimental curves. To illustrate this, consider two common equations used to describe the shear-thinning behaviour observed in viscometers. Figure 1.5 shows these two responses.

Figure 1.5a shows a steady shear-thinning response and the experimental points can be fitted to a simple equation:

$$\sigma = A_c\dot{\gamma}^n \tag{1.6}$$

where the two fitting parameters are A_c, the 'consistency', and n, the 'power law index'. This equation is often presented in its viscosity form:

$$\eta = A_c\dot{\gamma}^{n-1} \tag{1.7}$$

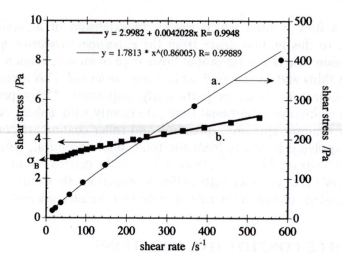

Figure 1.5 (a) *Power law behaviour from a 12% polyvinyl pyrrolidone solution;* (b) *Bingham plastic behaviour from a 14% w/v sodium kaolinite dispersion*

Figure 1.5b shows the behaviour of a 'Bingham plastic' and the fitting equation is:

$$\sigma = \eta_p \dot{\gamma} + \sigma_B \qquad (1.8)$$

Here the fitting parameters are the slope of the line (the plastic viscosity, η_p) and the Bingham or dynamic yield stress (the intercept, σ_B). Other constitutive equations will be introduced later in this volume as appropriate.

1.2.2 Using Constitutive Equations

The first use that we can make of our constitutive equations is to fit and smooth our data and so enable us to discuss experimental errors. However, in doing this we have the material parameters from the model. Of course it is these that we need to record on our data sheets, as they will enable us to reproduce the experimental curves and we will then be able to compare the values from batch to batch of a product or formulation. This ability to collapse more or less complicated curves down to a few numbers is of great value whether we are engaged in the production of, the application of, or research into materials.

The corollary is that we should always keep in mind the experimental range. Extrapolation outside that range is unwise. This will become particularly clear when we discuss the yield phenomenon – an area of great interest in many practical situations. Whatever the origins claimed for these models, they all really stem from the phenomenological study of our materials and so our choice of which one to use should be based on the maximum utility and simplicity for the job in hand.

1.3 DIMENSIONLESS GROUPS

An everyday task in our laboratories is to make measurements of some property as a function of one or more parameters and to express our data graphically, or more compactly as an algebraic equation. To understand the relationships that we are exploring, it is useful to express our data as quantities that do not change when the units of measurement change. This immediately enables us to 'scale' the response. Let us take as an example the effect of temperature on reaction rate. The well-known Arrhenius equation gives us the variation

$$k_r = A \exp(-E_a/RT) \qquad (1.9)$$

Here k_r is the rate of a reaction measured at temperature T, E_a is the activation energy and R is the gas constant. Now RT is the value of thermal energy and so the magnitude of the dimensionless group, (E_a/RT), immediately gives us a feel as to the importance of the activation process. For example, if $E_a \ll RT$, then the activation process will not slow the reaction rate significantly from the fastest possible rate A. On the other hand, if $E_a \gg RT$, then the reaction rate will be very much slower than A. Mechanistically this reminds us of the Boltzmann energy distribution and stochastic processes. The dimensionless group, (E_a/RT), is known as the 'Arrhenius Group'.

Another example from chemical kinetics can be seen in the rate equation for first-order reactions. Here the equation relating the concentration of a species A at time t, $[A](t)$, to the reaction time and the initial concentration, $[A](0)$, is

$$[A](t) = [A](0)\exp(-k_r t) \qquad (1.10)$$

The rate coefficient, k_r, has units of t^{-1} and so can equally well be thought of in terms of the characteristic time for the reaction to take place. Hence if $k_r t \gg 1$, the reaction will be a long way towards completion, whereas if $k_r t \ll 1$, very little change will have occurred. Equation 1.10 describes the decay of radioactive elements and $1/k_r$ could be considered as the characteristic time for the *relaxation* of the element from its active to its non-active state.

1.3.1 The Deborah Number

Maxwell introduced the idea of viscous flow as being the manifestation of the decay of elastically stored energy. If we follow this concept through we will see how a dimensionless group, the Deborah number, *De*, arises naturally. Let us consider a piece of matter in which all the molecules or particles (either small or large, it makes no difference) have had time to diffuse to some low energy state. Now if we *instantaneously* strain (deform) the material, we will store energy because the *structure* is perturbed and the molecules are in a higher energy state. As we hold the matter in this new shape, it becomes easier because the molecules diffuse until a low energy state equivalent to the initial one has been achieved, although the original shape has been lost, *i.e.* viscous flow has occurred. We can define the characteristic time it takes for this process to occur as the *stress relaxation time*, τ, of the material. Now the Deborah number is[3]

$$D_e = \frac{\tau}{t} \qquad (1.11)$$

The relaxation of the stress resulting from a step strain can be observed experimentally and we can see that it is the result of diffusive motion of the microstructural elements. Although we can have a mechanistic picture, what does this mean in terms of our measurements? We have the very striking result that our material classification must depend on the time *t*, *i.e.* the experimental or *observation time*. Hence, we can usefully classify material behaviour into *three* categories:

$$D_e \gg 1 \qquad D_e \sim O(1) \qquad D_e \ll 1$$
$$\text{solid-like} \qquad \text{viscoelastic} \qquad \text{liquid-like}$$

The most frequently quoted example to illustrate this behaviour is the children's toy 'Silly Putty', which is a poly(dimethyl siloxane) polymer. Pulled rapidly it shows brittle fracture like any solid but if pulled slowly it flows as a liquid. The relaxation time for this material is $\sim 1\,$s. After $t = 5\tau$ the stress will have fallen to 0.7% of its initial value so the material will have effectively 'forgotten' its original shape. That is, one could describe it as having a 'memory' of around $5\,$s (about that of a mackerel!). Many other materials in common use have relaxation times within an order of magnitude or so of $1\,$s. Examples are thickened detergents, personal care products and latex paints. This is of course no coincidence, and this timescale is frequently deliberately chosen by formulation adjustments. The reason is that it is in the middle of *our*, the human, timescale. Our nervous system responds over a timescale of 1 ms to 1 ks and so, if a material has a relaxation time within that region, we will observe an 'interesting' or useful texture. Reiner[3] pointed out that our observation time could be quite long with some materials, set concrete for example, and so ultimately our definition of solid-like can become one of practical rather than philosophic origin.

1.3.2 The Péclet Number

Although a mechanism for stress relaxation was described in Section 1.3.2, the Deborah number is purely based on experimental measurements, *i.e.* an observation of a bulk material behaviour. The Péclet number, however, is determined by the diffusivity of the microstructural elements, and is the dimensionless group given by the timescale for diffusive motion relative to that for convective or flow. The diffusion coefficient, D, is given by the Stokes–Einstein equation:

$$D = \frac{k_B T}{6\pi\eta_o a} \qquad (1.12)$$

where k_B is the Boltzmann constant, η_o is the viscosity of the liquid medium, and a is the radius of the diffusing moiety – molecule or particle. This has dimensions of $m^2 s^{-1}$. We can use Equation (1.12) to estimate the time taken for the diffusing moiety to move a characteristic distance. It makes sense to choose the radius as this distance and this gives us the Einstein–Smoluchowski equation:

$$t_a = \frac{a^2}{D} = \frac{6\pi\eta_o a}{k_B T}$$

and so

$$t_a = \frac{6\pi\eta_o a^3}{k_B T} \qquad (1.13)$$

Now the characteristic time for shear flow is the reciprocal of the shear rate. This is the time taken for a cubic element of material to be transformed to a parallelogram with angles of 45° (*i.e.* the time for unit strain to be applied) as shown in Figure 1.6. The Péclet number can now be written:

$$P_e = \frac{6\pi\eta_o a^3 \dot{\gamma}}{k_B T} \qquad (1.14)$$

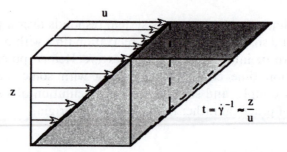

Figure 1.6 *The time taken for unit shear strain is $\dot{\gamma}^{-1} = z/u$*

An interesting problem arises when we consider solutions or colloidal sols where the diffusing component is much larger in size than the solute molecules. In dilute systems Equation (1.14) would give an adequate value of the Péclet number but not so when the system becomes concentrated, *i.e.* the system itself becomes a condensed phase. The interactions between the diffusing component slow the motion and, as we shall see in detail in Chapter 3, increase the viscosity. The appropriate dimensionless group should use the system viscosity and not that of the medium and now becomes

$$P_e = \frac{6\pi a^3 \sigma}{k_B T} \tag{1.15}$$

where the shear stress, $\sigma = \eta \dot{\gamma}$, has been used to make a clear distinction from Equation (1.14). Of course for a simple system, cyclohexane in decane for example, Equations (1.14) and (1.15) would give the same result as the intermolecular interactions between the species are similar and the viscosity of a mixture is similar to that of the two components. We shall use Equation (1.15) throughout as this indicates the importance of the interactions.

1.3.3 The Reduced Stress

The *reduced stress*, σ_r, was introduced by Krieger[4] from a dimensional analysis and has the form:

$$\sigma_r = \frac{a^3 \sigma}{k_B T} \tag{1.16}$$

The similarity to the Péclet number is obvious but we should also bear in mind the relationship to the Deborah number. This becomes clear when we consider the fact that the mechanism of stress relaxation is due to the diffusion of the microstructural components. For slow deformation processes – the low Deborah number, low Péclet number, or low reduced stress limit – the rate at which the structural elements can rearrange is great enough that the structure has little or no perturbation from that found in the quiescent state. Viscous deformation then occurs. Now if the straining is rapid, relaxation cannot take place and energy is stored. If the deformation is continuous, the structure must yield and breaking or 'melting' is then observed.

1.3.4 The Taylor Number

Common geometries used to make viscosity measurements over a range of shear rates are Couette, concentric cylinder, or cup and bob systems. The gap between the two cylinders is usually small so that a constant shear rate can be assumed at all points in the gap. When the liquid is in laminar flow, any small element of the liquid moves along lines of constant velocity known as streamlines. The translational velocity of the element is the same as that of the streamline at its centre. There is of course a velocity difference across the element equal to the shear rate and this shearing action means that there is a rotational or vorticity component to the flow field which is numerically equal to the shear rate/2. The geometry is shown in Figure 1.7.

When the shear rate reaches a critical value, secondary flows occur. In the concentric cylinder, a stable secondary flow is set up with a rotational axis perpendicular to both the shear gradient direction and the vorticity axis, *i.e.* a rotation occurs around a streamline. Thus a series of rolling toroidal flow patterns occur in the annulus of the Couette. This of course enhances the energy dissipation and we see an increase in the stress over what we might expect. The critical value of the angular velocity of the moving cylinder, Ω_c, gives the Taylor number:

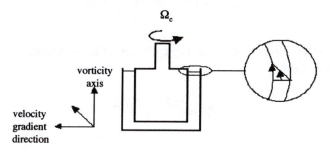

Figure 1.7 *Couette geometry*

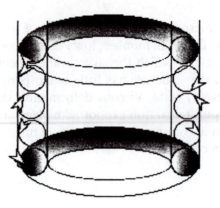

Figure 1.8 *Taylor vortices*

$$Ta = \Omega_c \left(\frac{R_o + R_i}{2} \right)^{1/2} \frac{\rho(R_o - R_i)^{3/2}}{2\eta} \tag{1.17}$$

where R_o and R_i are the outer and inner cylinder radii of the Couette filled with a fluid of density ρ and viscosity η. Figure 1.8 illustrates the flow pattern of Taylor vortices that are formed when the Taylor number is exceeded.

1.3.5 The Reynolds Number

The Taylor vortices described above are an example of *stable* secondary flows. At high shear rates the secondary flows become chaotic and turbulent flow occurs. This happens when the inertial forces exceed the viscous forces in the liquid. The Reynolds number gives the value of this ratio and in general is written in terms of the linear liquid velocity, u, the dimension of the shear gradient direction (the gap in a Couette or the radius of a pipe), the liquid density and the viscosity. For a Couette we have:

$$Re = \frac{\Omega_c(R_o + R_i)(R_o - R_i)\rho}{2\eta} \tag{1.18}$$

where R is the radius of the moving cylinder. When we write this in terms of the shear rate:

$$Re \approx \frac{\dot{\gamma}(R_o - R_i)^2 \rho}{\eta} \tag{1.19}$$

Another common geometry used for laboratory measurement of viscosity is a cone and plate with a small included angle, α. α is typically 1–5°. This geometry is used to give a constant shear rate because at any point on the plate the ratio of the tangential velocity ($r\Omega$) to the gap is constant. A suitable expression with the cone angle in degrees is

$$Re \approx \frac{\dot{\gamma}\rho}{\eta}\left(\frac{\pi R\alpha}{180}\right)^2 \tag{1.20}$$

In a tube we use the volumetric flow rate, Q, to calculate a mean velocity along the tube and we have

$$Re \approx \frac{Q\rho}{\pi R\eta} \tag{1.21}$$

It is important that we know at what Reynolds number our instrumental configurations give turbulent flow and work below this figure or we will think that shear thickening is occurring! A figure of $Re < 3000$ to $10,000$ is usually satisfactory for cone and plates or capillary viscometers, but values as low as 300 may be the maximum for some cup and bob units.

1.4 MACROMOLECULAR AND COLLOIDAL SYSTEMS

The range of diffusional timescales for dilute systems that are shown in Figure 1.9 and were calculated using Equation (1.13) immediately shows

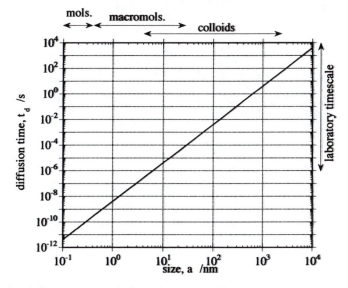

Figure 1.9 *Diffusional timescale from Equation (1.13)*

us that macromolecules and colloidal particles have diffusional time-scales that are within that of our laboratory instrumentation. Moreover as soon as we concentrate these systems, the motion of one unit is slowed by the interaction with its neighbours and timescales become much longer. Hence quite slow motions may be in the region of large Deborah or Péclet numbers. It is for this reason that polymers and clay particles are often added to formulations as 'thickeners' or 'viscosity modifiers'. We will deal with these systems in considerable detail in subsequent chapters and we shall see that it is their chemical nature, and the chemical environment that we put them in, which produces such a rich pattern of behaviour.

1.5   REFERENCES

1. R. Hooke, *Lectures "de Potentia Restitution"*, John Martyn, London, 1678, quoted in C.W. Macosko, *Rheology: Principles, Measurements and Applications*, Wiley-VCH, New York, 1994, p. 5.
2. I.S. Newton, *Principia Mathematica*, 1678, quoted in C.W. Macosko, *Rheology: Principles, Measurements and Applications*, Wiley-VCH, New York, 1994, p. 65.
3. M. Reiner, *Phys. Today* 1968, **17**, 62.
4. I.M. Krieger, *Advances Coll. Interface Sci.* 1972, **3**, 111.

CHAPTER 2

Elasticity: High Deborah Number Measurements

2.1 INTRODUCTION

In this chapter we will consider 'solid-like' behaviour. By this we mean that the stress relaxation time is long compared to the time taken to carry out our rheological observations. Thus τ may vary from milliseconds to such long times that we may take the value as infinite. So when we apply a stress to a material on a timescale where zero or only negligible relaxation can occur, the relative deformation or strain that we observe is recovered on the removal of the stress. The strain that we measure per unit stress that we apply is the *compliance* of the material. A highly compliant material is therefore one that exhibits a large strain for a small applied stress. On the other hand when we measure the stress resulting from a given strain, we use the stress per unit strain to give us the *rigidity* of the material and, of course, a very rigid material is one in which a small strain results in a large stress.

Systematic measurements of stress and strain can be made and the results plotted as a rheogram. If our material behaves in a simple manner – and it is surprising how many materials do, especially if the strains (or stresses) are not too large – we find a linear dependence of stress on strain and we say our material obeys Hooke's law, *i.e.* our material is Hookean. This statement implies that the material is isotropic and that the pressure in the material is uniform. This latter point will not worry us if our material is incompressible but can be important if this is not the case.

As an example, let us consider a typical response produced by stretching a sample of a vulcanised rubber. The components of the stress that we have to consider are the normal stresses σ_{xx}, σ_{yy} and σ_{zz}. Figure 2.1 illustrates the type of rheogram we get by plotting σ_{xx} against the strain. Note that the stress is calculated at each point as the

15

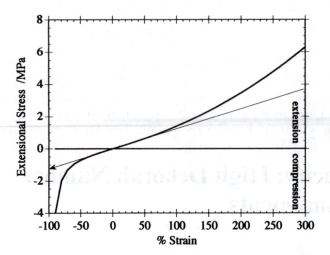

Figure 2.1 *Stress versus strain for a typical elastomer of vulcanised natural rubber. The straight line shows that a simple Hooke's law response would be observed up to ~40% strains*

extensional force divided by the cross-sectional area of the sample at that extension, whilst the percentage strain is the percentage elongation as compared to the original length of the sample.

The strains in Figure 2.1 are large but we can describe the response using one material constant, a *rigidity modulus*. This is termed a neo-Hookean model[1] and can give a reasonable description of elastomer behaviour up to large strains. (The convention that we are using is that extensional strain is positive whilst compression is negative, hence we require the modulus of the rigidity because the sign just indicates whether we are pushing or pulling.) Now we write the components of the stress resulting from the strain, ε, as:

$$\sigma_{xx} = G(1+\varepsilon)^2 - p$$

$$\sigma_{yy} = \sigma_{zz} = \frac{G}{1+\varepsilon} - p \qquad (2.1)$$

Here p is the pressure, which we are assuming is isotropic. We are also assuming that our material is incompressible, and the rigidity modulus, G, is what we would measure in a shear experiment. We can eliminate the pressure:

$$\sigma_{xx} - \sigma_{yy} = G\left(\frac{3\varepsilon + 3\varepsilon^2 + \varepsilon^3}{1+\varepsilon}\right) \qquad (2.2)$$

We are only pulling the ends of our block of elastomer and the sides are free, so $\sigma_{yy} = \sigma_{zz} = 0$. The curve shown in Figure 2.1 used a value for the rigidity modulus of 0.4 MPa.

If we work at small strains so that we are in the linear (Hooke's law) region of the rheogram, then Equation (2.2) reduces to

$$(\sigma_{xx})_{\varepsilon \to 0} = G3\varepsilon \tag{2.3}$$

Therefore the extensional rigidity modulus or Young's modulus can be written in terms of the shear modulus as

$$E = 3G \tag{2.4}$$

Generally our material will be compressible and $\varepsilon_{xx} \neq \varepsilon_{yy}/2 = \varepsilon_{zz}/2$ and we have to introduce another material parameter, Poisson's ratio, v. Poisson's ratio is defined as the ratio of the contractile to the tensile strain, *i.e.* $v = \varepsilon_{yy} : \varepsilon_{xx}$. Equation (2.4) now becomes:

$$G = \frac{E}{2(1 + v)} \tag{2.5}$$

So for an incompressible material $v = 0.5$ and Equation 2.4 is recovered. The value of Poisson's ratio for rubber is usually close to 0.5 but for many other solids the value is lower and we find $0.25 < v < 0.33$. We may also describe a Bulk rigidity modulus, K, such as we would measure when we compress a material with hydrostatic pressure, in terms of Young's modulus:

$$K = \frac{E}{3(1 - 2v)} \tag{2.6}$$

It is readily seen that when Poisson's ratio is 0.5, the material is incompressible because K becomes infinite.

2.2 THE LIQUID–SOLID TRANSITION

The simplest model for an atomic assembly is to consider the atoms as hard spheres with a radius a. Computer simulations have been used to describe the physical behaviour of such assemblies as the number density is changed. At low number densities the assembly is a fluid and the hard spheres diffuse in a gaseous fashion. There are three degrees of freedom corresponding to $k_B T/2$ for each orthogonal translational direction. At intermediate densities the motion of an individual sphere becomes more complex. Some of the time it will move inside a transient cage of neighbours and then move outside that cage. At the same time, when viewed on a larger scale, the local density fluctuates relative to the global average value. This means that we need to think in terms of three diffusion modes, *i.e.* diffusion coefficients: a short-time self-diffusion (motion within the cage), a long-time self-diffusion (a change of nearest neighbours), and a collective diffusion. The fluid assembly should now be thought of as a liquid. At higher number densities, the long-time self-diffusion mode disappears and we now have a solid. The internal energy, U, is given simply from the number density, N:

$$U = Nk_B T \qquad (2.7)$$

This is just the entropic contribution and the equations of state do not indicate a conventional liquid–solid transition. The computer simulations[2,3] however indicate that the transition occurs at a volume fraction of 0.495–0.54. The former figure is the freezing and the latter the melting density as the system is made to contract or expand. Rheologically this means that the viscosity of the system is increasing rapidly as the transition is approached and becomes infinite above this density. This is where the long-time self-diffusion coefficient has fallen to zero. However, the equations of state that describe the hard sphere assembly do not show a plateau in the pressure–volume curve, as we would expect in an assembly of van der Waals' atoms as it goes through a first-order phase transition from gas to liquid or liquid to solid. A second-order transition has zero entropy and volume change at the transition but this is not the case with a first-order system, where an increase in entropy accompanies an increase in volume on melting. 'Simple' or van der Waals' systems, argon for example, behave in this way. Water, however, decreases in volume on melting which is, of course, accompanied by an increase in entropy, due to the break-up of the very open ice structure.

When we consider a van der Waals' system, we can start with the pair interaction as shown in Figure 2.2. The equation giving the pair potential is the 6–12 or Lennard-Jones–Devonshire equation:

$$\frac{V(r)}{4V_m} = \left(\frac{2a}{r}\right)^{12} - \left(\frac{2a}{r}\right)^{6} \qquad (2.8)$$

Where $V(r)$ is the pair interaction energy and V_m is the well depth. Here $2a$ is the 'collision diameter', *i.e.* the distance at which the interatomic potential is zero. Also plotted in Figure 2.2 is the interatomic force and the elastic modulus of what we may consider as an interatomic spring. (Recall that the force is the rate of change of energy with distance,

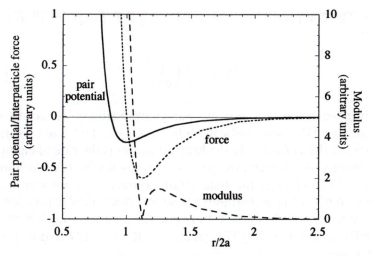

Figure 2.2 *Illustrative plot of the Lennard-Jones–Devonshire interatomic potential show-ing the force and the modulus curve for the pair interaction. Positive values indicate repulsion and negative values indicate attraction*

$-\partial V(r)/\partial r$, and the elastic modulus is the rate of change of force with distance, $\partial^2 V(r)/\partial r^2$.) When we deform a macroscopic sample of such a system the resistance comes from the sum of contributions from each of the atomic springs.

2.2.1 Bulk Elasticity

General physical chemistry texts[4] have made us familiar with the concept of the heat capacity being the rate of change of internal energy with temperature:

$$C_v = \left(\frac{\partial U}{\partial T}\right)_v \text{ and } C_p = \left(\frac{\partial H}{\partial T}\right)_p \tag{2.9}$$

where C_v is the heat capacity at constant volume and C_p is that at constant pressure; U is the internal energy and H the enthalpy $(H = U + pV)$. Why is this important to us here? A quick check of the physical chemistry texts gives us the compressibility, J_T, of a material in terms of the difference in heat capacities:

$$(C_p - C_v) = \frac{\alpha^2 T \bar{V}}{J_T} \tag{2.10}$$

where α is the coefficient of thermal expansion and $\bar{V}$ is the molar

volume. The compressibility is defined using the change in volume with
pressure:

$$J_T = -\frac{1}{V}\left(\frac{\partial V}{\partial p}\right) \tag{2.11}$$

Now, in rheological terminology, our compressibility J_T, is our bulk
compliance and the bulk elastic modulus $K = 1/J_T$. This is not a surprise
of course, as the difference in the heat capacities is the rate of change of
the pV term with temperature, and pressure is the bulk stress and the
relative volume change, the bulk strain. Immediately we can see the
relationship between the thermodynamic and rheological expressions. If,
for example, we use the equation of state for a perfect gas, substituting
$p\bar{V} = RT$ into $\alpha = 1/V(\partial V/\partial T)_p$ yields $\alpha = R/p\bar{V} = 1/T$ and so for our
perfect gas:

$$C_p - C_v = R \tag{2.12}$$

A plot of the change of C_v with temperature for glycerol is shown in
Figure 2.3. This shows how the heat capacity changes with temperature
and the nature of the phase present. As glycerol is cooled, the viscosity
rapidly increases with the result that a super-cooled liquid is produced
which forms a glassy solid. In the glass the degree of local order is high
but the long-range order that is apparent in a crystalline structure is
absent.

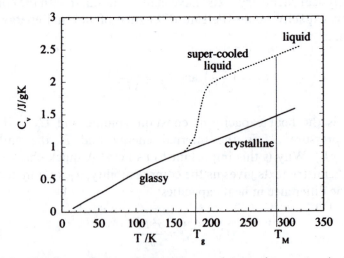

Figure 2.3 *The heat capacity of glycerol around its melting point (redrawn from
reference 5)*

If we consider the solid state and the pdV term (*i.e.* the value of $C_p - C_v$ at a given temperature) a little further but on an atomic scale, then we have two factors to consider. Firstly, we must be able to estimate the interatomic forces that are present, and secondly, we need to know the local structure if we are to model the effect of a volume change. To illustrate this we may consider crystals with a face-centred cubic structure. Three examples of this type of crystal are (1) van der Waals' solids, such as the inert gases, for which the pair potential is given by Equation (2.8), (2) metals with their delocalised electrons and (3) ionic crystals such as sodium chloride. In the crystals in an unstressed state the atoms/ions will be vibrating around the minimum in the potential energy (see, for example, the energy well in Figure 2.2). When a pressure p is applied we force the atoms/ions closer together and there is a volume change of $-pdV$. The simplest consideration is to make the work done in changing the internal energy due to a change in interatomic forces. Now we have $-pdV = dU$ and using Equation (2.11) we have:

$$K = \left(\frac{\partial^2 U}{\partial V^2}\right)V \tag{2.13}$$

As we know the structure and the dimensions of the atoms/ions it is relatively easy to write Equation (2.13) in terms of the energy minimum, V_m, *i.e.* the energy at the equilibrium separation. For the three cases we may write[6] the bulk modulus as:

$$K = B\frac{V_m}{V} \tag{2.14}$$

Where B is a constant for the type of atomic interaction and includes the Madelung constant:

$$B = -8 \text{ for the inert gases}$$
$$B = -2/9 \text{ for the alkali metals}$$
$$B = -1 \text{ for the alkali halides}$$

2.2.2 Wave Propagation

An important aspect of any structure is its ability to store energy when work has been done on it by imposing a strain. One result of this storage ability is that an impulse can be transmitted over significant distances, *i.e.* a wave can travel through the material over distances comparable to the wavelength or greater. Loss processes will inevitably occur and at low Deborah numbers the viscous processes rapidly damp out the oscillation

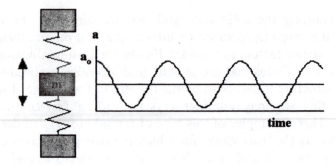

Figure 2.4 *Simple harmonic motion of the mass m caused by an initial displacement of a_0 from its equilibrium position*

close to the initiation point. The type of strain as well as the frequency (timescale) of the application is important. For example, a compression wave will travel considerable distances through water at sonic frequencies and provides a very effective communication mechanism over vast distances for marine mammals such as whales. However, a shear wave is damped very close to the moving surface.

Consider the situation shown in Figure 2.4 where a mass m is caused to oscillate by an initial displacement up to an amount a_0 at $t = 0$. The amplitude a would have to be smaller than shown for simple harmonic motion as a real spring would only obey Hooke's law over a limited strain amplitude. However the assumption is that Hooke's law is obeyed and the restoring force from both spring displacements is $-2ka_0$ where k is the force constant or elastic modulus of the spring. So we may write the force at any position as

$$m\frac{d^2a}{dt^2} = -2ka \tag{2.15}$$

i.e. $\dfrac{d^2a}{dt^2} = \dfrac{-2k}{m}a = -\omega^2 a$ if $\omega^2 = \dfrac{2k}{m}$

The solution to this equation can be written in terms of sines and cosines:

$$a = A\sin(\omega t) + B\cos(\omega t) \tag{2.16}$$

and the velocity of the motion as

$$v = \frac{da}{dt} = A\omega\cos(\omega t) - B\omega\sin(\omega t) \tag{2.17}$$

Now using the initial condition that $a = a_0$ at $t = 0$ in Equations (2.16) and (2.17) we have the following results:

$$a_0 = A \sin(0) + B \cos(0) \quad \text{and so } B = a_0$$
$$0 = A\omega \cos(0) - B\omega \sin(0) \quad \text{hence } A = 0$$

This gives the amplitude at time t as

$$a = a_0 \cos(\omega t) \tag{2.18}$$

The time taken for one complete oscillation is the period of the oscillation T_w:

$$T_w = \frac{2\pi}{\omega} = 2\pi \left(\frac{m}{2k}\right)^{1/2} \tag{2.19}$$

The above discussion will be familiar to all physical scientists, especially in the context of microwave spectroscopy and atomic vibrations. We should now consider what would happen if the mass m was in the middle of a long chain and the masses either side were free to move. As the mass moves, the force on the adjacent masses increases until $\omega t = \pi/2$. Following Tabor,[6] we can make the approximation that at this point the adjacent masses move in turn and a wave propagates through the material. This gives the wave velocity, v_w, as

$$v_w = \frac{4a_0}{T_w} \tag{2.20}$$

and then

$$v_w = \frac{4a_0}{2\pi} \left(\frac{2k}{m}\right)^{1/2} \tag{2.21}$$

and to a reasonable approximation for our purpose:

$$v_w^2 \approx a_0^2 \frac{k}{m} = \frac{a_0^3}{m} \frac{k}{a_0} \tag{2.22}$$

For a small compression/extensional strain $k/a_0 = E$, the Young's modulus of the system, and a_0^3 is the volume occupied by the mass m, and so using the density of the system, ρ, we have

$$E \approx v_w^2 \rho \tag{2.23}$$

Hence we can estimate the elastic modulus by measurement of the wave velocity and the density of a material. The simple treatment above can equally well be applied to a shear strain as a compressional strain with the result that

$$G = v_w^2 \rho \qquad (2.24)$$

where v_w is now the velocity of the shear wave. It should be noted here that, although a shear wave can readily be produced, the compressional wave used to give Equation (2.23) cannot be produced in a real material as any volume element of the material undergoes both a change in size and shape as the wave passes through. We should use rather the longitudinal bulk modulus,[7] M:

$$M = v_w^2 \rho \qquad (2.25)$$

where

$$M = K + \frac{4G}{3} \qquad (2.26)$$

and in terms of the Young's modulus and Poisson's ratio:[7]

$$K = \frac{E}{3(1 - 2v)}; \quad G = \frac{E}{2(1 + v)} \quad \text{so } M = E\left(\frac{1 - v}{(1 - 2v)(1 + v)}\right) \qquad (2.27)$$

(clearly for an incompressible material M is infinite).

2.3   CRYSTALLINE SOLIDS AT LARGE STRAINS

So far the discussion has really been concerned with small strains so that the elastic response can be considered to be linear. When the applied strain is increased steadily in a sample of a crystalline solid and the stress monitored, an interesting curve is produced. Figure 2.5 illustrates schematically the behaviour of a metal sample subjected to an extensional stress. An upper yield value is observed as the sharp peak where, at a point beyond the elastic limit, plastic flow occurs and permanent deformation has occurred. Beyond this the sample becomes thinner and the actual stress is higher due to the reduction in cross-sectional area. The corrected curve is shown as the broken line in Figure 2.5. This part of the curve shows the 'work hardening' region. If the elongation is now halted, the sample allowed to rest and then the strain progressively reapplied, we will observe an elastic response up to a level of stress slightly above that which the sample had previously reached. That is, the upper yield value will have shifted to a higher one just above the stress

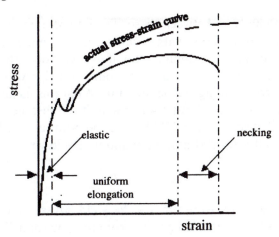

stress

actual stress-strain curve

elastic

necking

uniform
elongation

strain

Figure 2.5 *Schematic stress–strain curve for a metallic specimen such as wrought iron or mild steel*

which the sample had previously attained. This gives a larger elastic region but the modulus will be very similar to that found initially. To understand what has happened we need to consider what is occurring on a microstructural level.

2.3.1 Lattice Defects

Lattice defects often occur during crystal growth. Vacancies can occur, known as *Schottky defects*, when atoms or ions have migrated away from their lattice positions to the surface of the crystallite. When they leave a vacancy by migrating to a local interstitial position they are termed *Frenkel defects*. These types of defect weaken the crystal and reduce the modulus of the system. Perhaps more important than either of these defects is presence of dislocations where a mismatch in crystal planes occurs. These provide a source of weakness where plastic flow can occur, *i.e.* they define the limiting stress or upper yield point. Impurities can be incorporated in the structure, for example carbon and nitrogen can fit interstitially into the face-centred cubic iron lattice. When these impurities are strongly bonded into the lattice, they may restrict the movement along the dislocation and so increase the yield value markedly.

The importance of dislocations becomes evident when we consider the strain on the microstructure of a simple crystal. The atoms or ions in a crystal are in symmetric energy wells and so vibrate around their lattice site. When we track across a crystal plane, the potential energy increases and decreases in a regular fashion with the minima at the lattice points and the maxima halfway between these. The interatomic force, which is the rate of change of energy with distance, is at a maximum halfway between the maxima and minima. So if we visualise the energy as changing with distance along a line of atoms as a sine wave, the force would be the corresponding cosine function. What this means is that the maximum force occurs at about 1/4 of the unit cell distance and we can take this critical strain ($\gamma_{crit} = 25\%$) to be the point at which the plane would slip over into the next minimum and we could write:

$$\sigma_{y(max)} = \gamma_{crit}G \qquad (2.28)$$

In practice this grossly overestimates the yield stress, which may be a factor of 10^3 less than we would predict from this equation. The reason is that it is relatively easy for motion to occur across the end of the dislocation where there is a mismatch in the lattice planes. Of course the basic structure of the crystal is not changed and so when we pause the experiment and start again we find the same modulus. Figure 2.6 illustrates the process with a cubic lattice.

Work hardening of our crystal system occurs as the dislocations move and eventually meet and lock. However these are still potential weaknesses and increasing applied stress or strain will again produce motion. Eventually catastrophic failure occurs as the material fractures. Here dislocations can come together to produce a crack.

As a solid is produced by crystallisation from the melt, a space-filling multi-crystalline structure is produced. Each crystallite or grain has a different orientation from its neighbour and the mismatch of the lattices at the crystallite edges produces regions of poorer packing known as *grain boundaries*. Impurities can also be concentrated in these regions. During creep, the dislocations generated by plastic flow of the crystallites

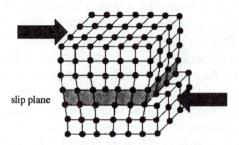

Figure 2.6 *A dislocation in a cubic lattice. As a shearing force is applied in the direction of the arrows, motion can most easily occur along the shaded plane, with the dislocation moving from left to right as the bonds 'flip' from one atom to the next*

can move through the grains to the grain boundaries at the same rate as they are being formed, *i.e.* a steady state is achieved. At higher deformation rates the dislocations build up and the grains become shattered and are reduced in size. At temperatures which are a significant fraction of the melting point, *i.e.* $T > 0.5T_m$, the creep rate, which in shear would mean the shear rate and in extension the extension rate, is governed by the diffusion of the atoms in the structure which enables the dislocation to remain unblocked.[6] The creep rate is then given by

$$\dot{\gamma} = A\sigma^n \exp\left(\frac{-E_D^*}{k_B T}\right) \tag{2.29}$$

where E_D^* is the activation energy for self-diffusion. Now the diffusion of an atom requires that there must first be an adjacent vacancy, so that the atom can then move into the space. We can visualise the process as that of the vacancy or hole migrating through the structure. The fraction of vacant sites can be estimated from the Boltzmann function and the energy required to remove an atom from its neighbours, ε_v,

$$\frac{n}{n_0} = \exp\left(-\frac{\varepsilon_v}{k_B T}\right) \tag{2.30}$$

where n_0 is the number of lattice sites. Due to the vacancy in the coordination shell, the energy ε_D required to cause an atom, adjacent to a vacancy, to jump into that vacancy is less than ε_v. v_m is the characteristic vibrational frequency of atoms in the lattice and this, when reduced by the Boltzmann factor, gives the number of potentially successful jump attempts:

$$\text{jump frequency} = v_m \exp\left(-\frac{\varepsilon_D}{k_B T}\right) \tag{2.31}$$

This is, of course, the reciprocal of the time taken for an atom to move a lattice site distance into a vacancy. v_m can be estimated from the heat capacity of the crystalline material using the Einstein or Debye models[6] of atoms as harmonic oscillators in a lattice. Combining Equations (2.30) and (2.31) gives the number of atoms moving per second as

$$\text{net jump frequency} = v_m n_0 \exp\left(-\frac{\varepsilon_D + \varepsilon_v}{k_B T}\right) \tag{2.32}$$

The diffusion constant, D, is given by the square of the distance moved per second and so

$$D = Pa^2 v_m n_0 \exp\left(-\frac{E_D^*}{k_B T}\right) \qquad (2.33)$$

because we can equate the activation energy for diffusion with that required to produce a vacancy and move a neighbouring atom into it. We have included a constant P in the equation to recognise that this would only give an estimate because details of the motion of the surrounding atoms should really be included, as should the directional probability of any vibration. However it is reasonable to expect P to be of the order of 1.

Annealing the system at temperatures close to its softening point allows recrystallisation to occur and the grain size to increase. This process again progresses by diffusion of holes through the structure and it is quite clear from Equation (2.33) that this process will be assisted by elevated temperatures.

2.4 MACROMOLECULAR SOLIDS

This is a very important class of materials which includes both natural materials such as wood, teeth and opals and high technology materials such as synthetic rubbers and other composite materials. We will however split this category into polymers and particulate systems, because different modelling is required for these, although both groups are made up of units consisting of very large numbers of atoms. The topic of polymers will be addressed below and particulate systems will be discussed in Section 2.5.

2.4.1 Polymers – An Introduction

There are three situations that appear to be relevant here. First, we may think of a solid polymer formed from the melt; second, the much more compliant elastomers that initially come to mind when we think of rubber elasticity; and third, polymer gels formed in polymer solutions. In each case the details of the physical chemistry of the macromolecules is crucial to the understanding of the structure that is formed. In this section we will concentrate on organic macromolecules because the rheology of these molecular systems is often the reason for their use.

The backbones of polymers are made up of carbon chains. The bonds are usually single but there may be a significant number of double bonds. The side groups or side chains are a key feature as it is these that control the rotation around the bonds and it is the bond rotation which gives a polymer molecule its flexibility. The covalent C—C bond is rigid and the bond angle is fixed but, as there are a great many bonds making up the backbone, rotation allows many different coil shapes to be formed as well as a relatively easy change from one conformation to another. The result is a system that is much more compliant in shear and extension than other types of solids (*i.e.* metals, ionic or covalent). The Poisson's ratio is usually close to 0.5 and from Equation (2.27) we can see that as

$$K = G \frac{2(1 + v)}{3(1 - 2v)} \tag{2.34}$$

the bulk modulus can be large.

2.4.2 Chain Conformation

Prior to a discussion of the theory of rubber elasticity, it is important to review how isolated polymer chains behave as this will provide a picture of the size and shape of a polymer. Clearly a polymer chain in a vacuum will collapse into a dense unit, but when in a solution the molecule will take on a conformation which is a function of the interaction with the surrounding molecules and the balance between the entropically driven tendency to maximise the spatial configuration and the connectivity of the monomer units. This is the case whether the chain is surrounded by small molecules (solvent) or other macromolecules that may or may not act like a solvent.

In a *dilute* polymer solution the chains are separated and take up some coil-like conformation. The dimensions can be estimated from the statistics of a random walk. Of course, the coil is not static and on average the shape is ellipsoidal. Detailed descriptions of the statistical conformation of polymer chains in solution is given in the literature[8,9,10] and here we will just summarise the simple results. When we do this it must also be remembered that the connectivity of the chain components means that polymer molecules and solvents are necessarily non-ideal. The molecular weights are markedly different and all the behaviour of polymers in solution occurs at very small mole fractions even though the weight or volume fractions are high. The molecular weight of a chain is the monomer weight, m_m, multiplied by the degree of polymerisation, X, *i.e.* it is simply $m_m X$. The monomer units may be linear hydrocarbon chains, with or without side groups, or they may consist of larger, less mobile units, as in cellulose where we must think of the basic unit as being a glucose ring that will act as a single link.

It has already been pointed out that the rotation about the chemical bonds linking the building blocks of the chains allows many possible conformations. The simplest model of a polymer chain is that of a 3-D

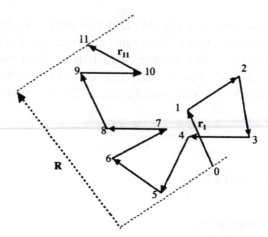

Figure 2.7 *Simple random walk in 3-D showing equal magnitude steps r_1 to r_{11}. The end-to-end vector is* **R**

random walk with fixed step lengths, *i.e.* the problem of fixed bond angles and physical atomic volume is ignored. Figure 2.7 shows such a model. The important parameter that we need to characterise this walk is the magnitude of the vector from atom 0 to atom 11, shown as **R** in Figure 2.7. Taking our origin as the position occupied by atom 0, the value of **R** is simply the sum of all the individual bond vectors:

$$\mathbf{R} = \sum_{n=0 \to 11} \mathbf{r}_n \tag{2.35}$$

The problem is that we must calculate an average value of this end-to-end length. There will be as many bonds going in one direction (to the right being defined as the positive direction) as in the opposite (to the left being defined as negative). Therefore the mean value of the vector will be zero. However, we can solve the problem of the directionality by using the square of each and then taking the square root of the average value. Hence we define the end-to-end dimension as the root mean square value of N bonds of length b:

$$\bar{R} \equiv \langle \mathbf{R}^2 \rangle^{1/2} = b\sqrt{N} \tag{2.36}$$

Figure 2.8 shows part of a polyethylene glycol chain and illustrates the type of problem encountered with real chains. First, there are fixed bond angles (in this case the tetrahedral angle), and second, there is the physical volume occupied by each atom so that the walk cannot for example cross itself. Both of these constraints mean that the random

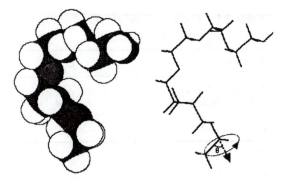

Figure 2.8   *Part of a polyethylene oxide chain using wire frame and space-filling models. Rotation about a carbon atom means that the following bond sweeps out a cone with an included angle of 2θ. The hydrogen atoms are white and the oxygen atoms are grey*

walk is more open than the simple model. We can express this expansion by introducing a factor, C_∞ , known as the characteristic ratio:

$$\bar{R}^2 = C_\infty N b^2 \qquad (2.37)$$

$C_\infty b^2$ is the 'effective bond length', b'. For the case of the bond angle restriction on the random walk:

$$C_\infty = \frac{1 + \cos \theta}{1 - \cos \theta} \qquad (2.38)$$

and so for the tetrahedral angle $\theta = 180° - 109.5°$ and $C_\infty = 2$. Experimentally we can think of this characteristic ratio as describing the coil expansion due to constraints imposed on the random walk such as fixed bond angles and restricted rotation due to bulky side groups, and exclude volume effects. The coil size could be determined experimentally, for example from a viscosity study. Alternatively light or neutron scattering may be used to characterise the polymer coil and here the radius of gyration, R_g, is determined. R_g is the average distance of each atom from the centre of mass of the coil and is directly related to the root mean square dimension by

$$R_g = \frac{\bar{R}}{\sqrt{6}} \qquad (2.39)$$

Table 2.1 illustrates the magnitudes of the characteristic ratio found for typical polymers in dilute solution. The relatively simple poly(ethylene oxide), (PEO), chain is fairly flexible whereas the cellulosic chain has

Table 2.1 *Values of the characteristic ratio for various chains*

Polymer	C_∞
Unhindered rotation	2.0
Poly(ethylene oxide)	3.4
Cellulose ether	8.0
Poly(styrene)	10.0

much more rigid glucose rings and the poly(styrene) chain has large side groups on every second carbon atom in the backbone.

So far the size and conformation of the polymer molecule has only been considered in terms of the chemical architecture and the statistics of a random walk. The chemical environment of the chain is also important, *i.e.* the solvent and the temperature that was used to prepare the solution. This again focuses attention on the interaction energies between molecules. Formally it is the free energy of the interaction between a solvent molecule and a chain segment compared to both that acting between two chain segments and that acting between two solvent molecules. The simplest theory is the Flory–Huggins theory[11] which is based on a simple cubic lattice model. The entropy is calculated from the number of solvent–segment interactions while the enthalpy is calculated from the difference in interaction energy between the various combinations of segment and solvent molecular interactions. The use of a simple cubic lattice is not an important restriction, but bigger problems arise from the assumption that solvent molecules and polymer segments are the same size. The free energy of mixing estimated from this model is

$$\Delta G_m = k_B T [n_1 \ln \varphi_2 + n_2 \ln \varphi_2 + n_1 \varphi_2 \chi] \qquad (2.40)$$

where n_i is the number of lattice sites per unit volume occupied by solvent ($i = 1$) or polymer ($i = 2$) and the probability of a site being occupied is proportional to the relative volume, *i.e.* to $\ln \varphi_i$. The first two terms in the bracket come from the number of ways the polymer and solvent can be arranged and these are taken as the entropic contribution. The third term accounts for the free energy of the contact between polymer and solvent. This is assumed in the original theory to be the result of the enthalpic contribution and so the χ-parameter is

$$\chi = \frac{\Delta H_m}{n_1 \varphi_2 k_B T} \qquad (2.41)$$

As we have the free energy of mixing, we may now estimate the osmotic pressure of our *dilute* polymer solution:

$$\Pi = RT\left[\frac{c_2}{M_2} + \left(\frac{\bar{v}_2}{M_2}\right)^2 \frac{1}{\bar{v}_1}\left(\frac{1}{2} - \chi\right)c_2^2 + \ldots\right] \quad (2.42)$$

$\bar{v}_i$ is the molar volume of polymer or solvent, as appropriate, and the concentration is in mass per unit volume. It can be seen from Equation (2.42) that the interaction term changes with the square of the polymer concentration but more importantly for our discussion is the implications of the value of χ. When $\chi = 0.5$ we are left with the van't Hoff expression which describes the osmotic pressure of an ideal polymer solution. A solvent/temperature condition that yields this result is known as the θ-condition. For example, the θ-temperature for poly(styrene) in cyclohexane is 311.5 K. At this temperature, the poly(styrene) molecule is at its closest to a random coil configuration because its conformation is unperturbed by specific solvent effects. If χ is greater than 0.5 we have a poor solvent for our polymer and the coil will collapse. At χ values less than 0.5 we have the polymer in a good solvent and the conformation will be expanded in order to pack as many solvent molecules around each chain segment as possible. A θ-condition is often used when determining the molecular weight of a polymer by measurement of the concentration dependence of viscosity, for example, but solution polymers are invariably used in better than θ-conditions.

It is now useful to consider how we may define the concentration of polymers in solution. Dilute solutions are ones in which the polymer molecules have space to move independently of each other, *i.e.* the volume available to a molecule is in excess of the excluded volume of that molecule. Once the concentration reaches a value where there are too many molecules in solution for this to be possible, the solution is considered to be *semi-dilute*.[8] The concentration at which this occurs is denoted by c^*. For convenience c^* is defined as the concentration when the volumes of the coils just occupy the total volume of the system, *i.e.*

$$c^* \approx \frac{M}{N_A}\frac{3}{4\pi R_g^3} \quad (2.43)$$

At concentrations greater than this value there will be polymer chains at any position in the solution but there will still be large fluctuations in local concentration with position because the density of each polymer coil is greatest at the centre of mass of the coil. The interactions between segments of neighbouring coils due to overlap progressively screen out the intramolecular segment interactions and the coils move towards a more ideal conformation. As the concentration is increased, the local fluctuations become less and the local concentration can be taken as the average over the system. The polymer solution is now known as *concentrated* and the concentration where we can assume a uniform polymer density is c^{**} where[8]

$$c^{**} \approx \frac{v}{b'^6} \frac{M_s}{N_A} \qquad (2.44)$$

Here v is the excluded volume of a chain segment, b' is the bond length of that segment with a molar mass of M_s. In order to give the rotational flexibility to produce a random walk a chain segment of bond length $b' = C_\infty b$ is used, representing a group of monomer units. The excluded volume is of course related to the χ-parameter and may be written[12]

$$v = \frac{\tilde{v} M_s}{N_A} (1 - 2\chi) \qquad (2.45)$$

where $\tilde{v}$ and is the specific volume of the polymer. Clearly as the solvent quality improves, the excluded volume increases and the chains expand. As the concentration rises above c^{**}, we approach the melt.

2.4.3 Polymer Crytallinity

At this stage we only have part of the picture because we must think about what happens when the chain density is high. When formed from a melt most solid polymers have a degree of crystallinity and this has a marked effect on their properties. The crystalline regions consist of sections of the polymer chains that are closely aligned and so maximise the van der Waals' interactions. During cooling of the melt, these regions nucleate, as we would expect with any crystallisation process, but because of the connectivity provided by the long macromolecular chains, these regions are quite small and one molecule can be associated with several crystallites. This is shown in Figure 2.9. Hydrocarbons with a low molecular weight also form crystals due to London–van der Waals' attractions. The C_{20} to C_{30} paraffinic hydrocarbons form a wide range of useful crystalline waxes. In these cases, however, plastic flow readily occurs because the crystals are not interconnected, as in the case of polymers.

Processing of polymers during the solidification process can be used to control the crystallisation, so for example when a high extensional

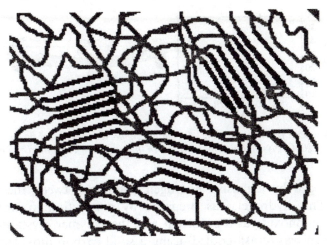

Figure 2.9 *Schematic of a linear polymer cooled from the melt showing crystallinity as the darker regions of parallel sheets of chains folded back and forth*

deformation is produced (fibre spinning) the crystallinity can be enhanced, but more importantly for many applications, it can be aligned with the fibre axis. This provides maximum strength with minimum compliance. Sheet polymer can also be aligned by drawing. Processing to produce maximum strength by crystal manipulation has long been used with metals, of course, with wrought iron being an obvious example. Although the very high tensile strength is important (linearised poly (ethylene) fibres have a much greater tensile strength than a similar weight steel), we should also note the low compliance. This is because in the crystals the molecules are stretched out and the drawing process also tends to stretch out the connecting molecules as well so that there is little coil structure left to unravel. The point here is that the compliance of a polymer solidified from the melt is a combination of that of the crystallites with the much more compliant, or 'spring-like', interconnecting chains. In other words, if we look at the deformation produced by a load, it is the response of the interconnections that we are observing. Systems of this type can form high modulus solids. The modulus can be measured by simple means such as bending or stretching a sample. Of course care must be taken at any clamp points to avoid the introduction of errors due to stress concentration at corners, sample deformation, *etc*. An easy and simple experiment is to study the bending of a simply supported beam with increasing load. A suitable geometric arrangement is depicted in Figure 2.10. The experimental data shown in the figure is from a hydrogel. This was prepared using poly(vinyl alcohol), PVA. PVA is difficult to dissolve in water due to its high degree of crystallinity, which

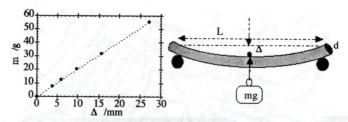

Figure 2.10 *Data from the bending of a simply supported cylindrical beam consisting of 20% polyvinyl alcohol in water. In this case L = 110 mm; d = 7 mm*

is encouraged by extensive hydrogen bonding between the hydroxyl groups. This hydrogel was prepared by dissolution in water using dimethyl sulphoxide as a cosolvent and the polymer crystallised when the solution was 'crash-cooled' using a solid carbon dioxide/methanol slush bath.[13] Sheets, tubes and rods can be produced in this way. When a beam is bent in the manner illustrated in Figure 2.10, the upper and lower halves carry the maximum compression and tension loads respectively. But there are also shear components in the beam, *i.e.* there are components of compression and tension acting diagonally (at 45° to the beam axis but normal to each other). The expression for the Young's modulus is given by[14]

$$E = \frac{mg}{\Delta} \frac{L^3}{48I} \qquad (2.46)$$

where Δ is the deflection caused by the applied load mg, L is the length of the hydrogel rod between the supports, d is the rod diameter and I is the moment of inertia of the section, which for a circular cross-section is

$$I = \frac{\pi(\frac{d}{2})^4}{4} \qquad (2.47)$$

Hence Equation (2.46) can be written

$$E = \frac{mg}{\Delta} \frac{4}{3\pi d} \left(\frac{L}{d}\right)^3 \qquad (2.48)$$

The data for a hydrogel at 20% polymer is shown in Figure 2.10 and gives a value of the Young's modulus $E = 11.8$ MPa.

It should be noted that this is an equilibrium value that was not reached immediately. The initial value is higher and falls with time to this value. Hydrogels of this type are a little different from a polymer prepared from a melt in that they are heavily swollen with a liquid

(80% water in this case). When the load is initially applied the gel is incompressible until the fluid starts to drain out through the mesh. In the initial stage it is the shearing that controls the bending, *i.e.* the shear modulus can be calculated at this stage from the deflection. Scherer[15] made use of this by applying a deformation and measuring the force as a function of time with silica gels that were covalently linked, however the technique is applicable to other systems. If the initial load is $W(0)$ and the final load is $W(\infty)$, [equivalent to mg in Equation (2.46)], we have the pair of equations:

$$W(0) = \frac{144I\Delta}{L^3}G \tag{2.49a}$$

$$W(\infty) = \frac{48I\Delta}{L^3}E \tag{2.49b}$$

The ratio of Equations (2.49b) to (2.49a) and Equation (2.5) provides a measure of Poisson's ratio as

$$\frac{W(\infty)}{W(0)} = \frac{E}{3G} \quad \text{and so} \quad v = \frac{3W(\infty)}{2W(0)} - 1 \tag{2.50}$$

For a network of springs the shear modulus is[16,17,18]

$$G = n_s k_B T \tag{2.51}$$

where n_s is the spring density in the system. This may be written in terms of the molecular weight of the polymer springs as

$$G = c_p \frac{N_{Av}}{M_s} k_B T \tag{2.52}$$

where c_p is the polymer concentration, N_{Av} is the Avogadro Number, and M_s is the molecular weight of the spring. Using a value of Poisson's ratio of 0.3, we have a value of $M_s = 110$ for our 20% PVA hydrogel. This gives an *average* of two or three monomer units between the 'crystallites' and indicates the high degree of organisation in the system. This is obvious from the high value of the modulus, which is two or three orders of magnitude stiffer than a food grade gelatin gel. The fact that the PVA hydrogels are transparent is also important because this shows clearly that the 'crystallites' must be very small, otherwise they would scatter light and an opaque gel would result.

2.4.4   Crosslinked Elastomers

The picture can now be extended to elastomers where the polymer chains have an occasional covalent crosslink tying them together. This gives a structure that consists entirely of interconnected springs because there are no crystallites present. When this structure is deformed the conformation of the polymer coils is constrained causing a decrease in entropy, *i.e.* the 'springs' are 'entropic springs'. This has the interesting result that the elastic modulus *increases* with an increase in temperature in contrast to what we observe with other solids. Swelling of the network with a solvent reduces the elastic modulus by reducing the number of 'springs' in a unit volume. It will not change the crosslinking as these are covalent bonds. Without the crosslinking the system would no longer be a solid. It would be a liquid, albeit a viscoelastic liquid, with a finite viscosity at low shear rates or low oscillation frequency, *i.e.* it would have a zero shear viscosity. What happens in this case is that polymer molecules, which of course are in constant motion, can 'wriggle' through the mutual structure and so maintain a conformation that corresponds to a low energy state. This motion is called *reptation*. At higher deformation frequencies the diffusive motion of the polymer molecules is too slow to free all the entanglements and we start to see an elastic response. At sufficiently high frequencies, none of the entanglements will have sufficient time to relax the stress by reptation and all will be acting as effective crosslinks. The elastic modulus reaches a plateau value at frequencies in excess of this level and the network modulus, G_N, has been reached. Polymer melts and solutions of high molecular weight polymers behave in this manner and form what are called *physical gels*. This behaviour of the storage modulus as a function of frequency is illustrated schematically in Figure 2.11. The network modulus can be expressed from Equation 2.51 in terms of the polymer concentration (mass per unit volume), c, and the molecular weight between the entanglements:[8]

$$G_N = c \frac{N_A}{M_e} k_B T \qquad (2.53)$$

Equation (2.53) is stating that the network modulus is the product of the thermal energy and the number of springs 'trapped' by the entanglements. This is the result that is predicted for covalently crosslinked elastomers from the theory of rubber elasticity that will be discussed in a little more detail below. However, what we should focus on here is that there is a range of frequencies over which a polymer melt behaves as a crosslinked three-dimensional mesh. At low frequencies entanglements

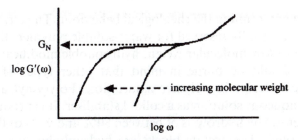

Figure 2.11 *Schematic of the storage modulus as a function of the frequency of the applied strain for a polymer melt. The plateau gives the network modulus, G_N. The plateau extends down to lower frequencies as the molecular weight is increased because the relaxation time is proportional to M^3*

will not hold as the reptation motion of the chains allows relaxation of the stress – due to the work stored when distorting the chain – and at very low frequencies all of the polymer molecules can diffuse rapidly enough to maintain the preferred coil-like conformation. As the molecular weight is increased, this centre of mass diffusion becomes slower and slower and so the experiments must be carried out more slowly to see this relaxation. These techniques are discussed in detail in later chapters. At this stage we will assume that the slowest relaxation time is much less than the experimental time. At very high frequencies the modulus increases above the G_N value as the experimental timescale becomes so short that only the motion of shorter and shorter sections of the chain can be accessed. This means that the mesh appears experimentally more and more like a rigid unit as the frequency is increased until we are in effect just looking at the stiffness of the C—C bonds making up the network.

2.4.5 Self-associating Polymers

Many polymers are not just simple homopolymer chains but are in fact *copolymers*. They may be synthesised so that there are large blocks of a second or even a second and third polymer species along the chain. The latter are known as *terpolymers*. With these systems the detailed chemistry is of great importance. This is very clearly demonstrated by looking at the properties in aqueous solution. The reason is that water is a strongly hydrogen-bonding solvent and will interact differently with different types of polymer. Some parts of the chain may be less soluble than others because they cannot fit into the water structure and so they tend to cluster together. To illustrate this we will consider in some detail the polymers used as 'associative thickeners'. These are particularly good examples of how the chemical details of a polymer chain are of major

importance in determining the rheological behaviour. The term 'associative thickener' is usually reserved for water-soluble polymers that have a
small amount of low molecular weight hydrophobic modification to the
chain but it should be borne in mind that other types of chemical
architecture can result in polymer association. Poly(vinyl alcohol) is
often used in aqueous solution as a colloid stabiliser. It is produced from
poly(vinyl acetate) by hydrolysis of between 80% and 98% of the acetate
groups to alcohol. The acetate blocks are hydrophobic and provide a
gelling mechanism by association of the hydrophobic regions. Multi-
valent ion bridging is another type of gelling mechanism that binds
chains together. Calcium ions can be used to gel poly(acrylic acid) or
cobalt ions to gel hydroxypropyl guar.

For a polymer to dissolve in water, the high molecular weight chain
must be capable of 'fitting into' the hydrogen-bonding structure. Chemical groups such as -OH or -COOH can readily do this whilst aliphatic or
aromatic hydrocarbons cannot. Hence around a simple hydrocarbon
chain there is a discontinuity in the water structure and we may visualise
the hydrocarbon chain as being in a 'cage'. When there are many of these
in a given water volume a significant gain in free energy can be achieved if
the hydrocarbons are grouped together to give a larger cage. Clearly
there is a balance here between the internal energy advantage of
increasing the hydrogen bonding as well as increasing the interaction
between hydrocarbons and the decrease in entropy due to the hydrocarbon clustering. However there is a net advantage in the arrangement
of $\sim k_B T$ per methylene group and this is the origin of the hydrophobic
bond.

Short hydrocarbon chains with 12 to 18 methylene groups may be
readily attached to water-soluble polymer chains. Only a few hydrophobes are required on each polymer to produce a useful thickener. High
degrees of addition produce very strong gels, *i.e.* ones with a very high
modulus. Polymers of this type are generally classed as hydrophobically
modified and the modification of the water-soluble backbone may be
done in a number of ways. The chain may be terminated at each end by
hydrophobes. Poly(ethylene oxides), terminated with an octadecyl chain,
for example through a urethane link, are known as HEURs.[19] Cellulose
ethers with three or four random grafts of, for example, hexadecyl chains
are known as HMHEC[19] (the etherification of the cellulose is required to
render the chain soluble by reducing the ability to crystallise). Poly
(acrylamides) can be synthesised to give a few small bunches of hydrocarbon chains along the main backbone and can then be labelled
HPAMs.[19]

These types of polymers have some particular advantages as thickeners as they can be useful at relatively low molecular weights. This is important because it reduces the elastic responses during flow, *i.e.* energy storage is less as relaxation is rapid. A practical consequence of this is that low spatter coatings can be manufactured. However, under quiescent conditions, a three-dimensional network is produced by hydrophobic bonding. This is weak and readily broken by shear or extensional strains but is self-healing when deformation ceases.

Each of the examples mentioned above behave slightly differently and these differences are due to the detailed structure. In each case the hydrocarbon groups associate via hydrophobic bonding but HMHEC for example can show a critical concentration threshold for this to occur. HEUR on the other hand tends to associate at all concentrations. This is due to accessibility of the hydrophobes as they are at the ends of very flexible chains. In HMHEC, however, we have a much stiffer chain with the hydrophobes spread randomly along it. It is therefore a more difficult process to bring these together to form network nodes. HPAM conforms more closely to HMHEC than HEUR but, as we have groups of hydrocarbon chains at each modification site, it associates somewhat more readily.

In order to clarify this behaviour further, we will now discuss an HEUR in more detail using a specific example that was prepared from a monodisperse poly(ethylene glycol) with a number average molar mass of 35×10^3 Dalton. The terminal hydroxyls were capped by octadecyl chains.[20] In the dilute regime aggregation of the hydrophobes occurs by a mixture of inter- and intra-molecular associations. c^* is reached at $73\,\mathrm{kg\,m^{-3}}$. Below c^* there is a greater probability of intra-molecular interactions resulting in 'closed loop' network defects. This reduces the number of elastically effective network links in the gel as each end of the closed loop is connected to the same node. Above c^* adjacent polymer coils are overlapping and the probability of this type of network defect is reduced. At concentrations where $c \gg c^*$ an increasing degree of chain entanglement is to be expected. These will be effectively 'locked in' by the chain ends being in nodes formed by the clusters of hydrophobes. Thus the number of elastically effective links increase with each entanglement. The chain entanglement is a bimolecular process and therefore gives a c^2 contribution to the modulus of the gel. The remaining type of network defect that could occur is an unassociated hydrophobe that, like the closed loop case, reduces the number of elastically effective links, in this case by one link for each free hydrophobe site.

Now for our example of an HEUR gel the concentration used was $52.5\,\mathrm{kg\,m^{-3}}$ so that we had a value of $c/c^* = 0.72$. This means that we may consider entanglements to be unlikely but that some closed loops are likely to have been formed. This was confirmed experimentally by the value of the network modulus, $G_{N\,expt}$, being linear with polymer concentration indicating no c^2 dependence. The way that the network modulus was measured experimentally was by oscillation of the gel at frequencies much higher than the reciprocal of the slowest relaxation time. The slowest relaxation process involved the removal of a hydrophobe from a cluster and for this gel was $\tau \sim 8\,s$ with $G_{N\,expt} = 2.6\,kPa$. Now in an ideal network of this type of polymer each chain could be expected to form an elastic link. The number of chains per unit volume is

$$n_c = \frac{cN_{Av}}{M_n} \tag{2.54}$$

and so we calculate the network modulus as

$$G_{N\,calc} = \frac{cRT}{M_n} \tag{2.55}$$

where R is the gas constant. Hence for this system we obtained:

$$\frac{G_{N\,expt}}{G_{N\,calc}} = 0.7 \tag{2.56}$$

which indicates that 30% of the polymer chains are ineffective as elastic links. The aggregation number of the hydrophobes in the clusters has been shown to be $N_{agg} \sim 6$,[21] but the value may be as high as 20. This is much lower than that found for the micellar clusters of common surface active agents (surfactants) such as sodium dodecyl sulphate, SDS. The latter form spherical clusters at low concentrations containing ~ 74 hydrophobic chains.[22] When surfactants are added to an HEUR solution, mixed clusters are formed and with SDS, $N_{agg} \approx 110$[23] which means that ellipsoidal micellar clusters are formed. This is to be expected, due to the mismatch in the lengths of the two types of hydrophobe involved.

The situation with the HMHEC polymers is a little more complicated and we can use their behaviour to illustrate the importance of network defects. There are a small number of hydrophobes randomly grafted along a relatively stiff chain and the hydroxyethyl cellulose is prepared by the degradation of a naturally occurring polymer so the molecular weight distribution is broad. However, rheological measurements are an integration over a very large number of interactions and so give a good measure of the average. Also the hydrophobes are of a narrow size distribution and these dominate the stress relaxation times. The aggregation number of the hydrophobe cluster is important in defining the range of relaxation times that we see. If the aggregation number is large (say $N_{agg} > 10$), the local environment of any hydrophobe will vary little with the actual aggregation number occurring at any given time for its cluster. That is to say that the energy required to enable it to escape from its cluster becomes only a weak function of N_{agg} and this leads to a sharp relaxation response. If the aggregation number is low, as occurs with HMHEC due to a combination of chain stiffness and steric restrictions, the local variations lead to a broadening of the relaxation response. In terms of this chapter it is only important in setting the high frequency value that we choose for our experiment. So we have to be sure that there is little or no relaxation of the network and our Deborah number is sufficiently high that a reliable value of G_N, our network modulus, is obtained. The data shown in Figure 2.12 were obtained by wave propagation experiments[24] where the frequency is $\omega \sim 1.2 \times 10^3 \, \text{rad s}^{-1}$. The curvature of the data is a clear indication that 'locked in' chain entanglements are contributing to the value of the network modulus. We will now consider how we can model this type of transient network. The system used had an $M_n = 1 \times 10^5$ Dalton and there were an average of ~ 4 hexadecyl chains per hydroxyethyl cellulose chain in the synthetic

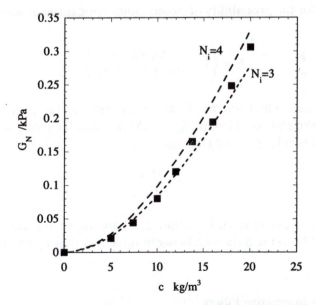

Figure 2.12 *The network modulus of a solution of an HMHEC in water, showing the curves calculated from Equation (2.58) and the experimental points taken from reference 24. The molecular weight was $M_n = 10^5$ and there were ~ 4 hexadecyl hydrophobes per cellulose chain at synthesis*

mixture so that the maximum number of crosslinking sites per chain, N_i, was ~4.

If $N_i = 1$, when chains are linked together a network will not be formed but just longer chains. When there are two or more, each site can lead to a network spring if more than two hydrophobes can cluster to form a node. (Note that this is different from the situation of a chemically crosslinked network where only two chains would be joined by the formation of the link. In this latter case a value of $N_i > 2$ is required for a network.) The potential number of network springs is

$$n_s = \frac{(N_i - 1)}{2} \frac{N_{Av}c}{M_n} \tag{2.57}$$

Here the 2 in the denominator is to avoid double counting because it takes two sites to form one link. However closed loops and chain entanglements are both possibilities and Equation (2.57) must be modified for these effects. We can write the network modulus as

$$G_N = n_s k_B T(1 - f_1) + Bc^2 \tag{2.58}$$

where the entanglement correction coefficient was assigned a value of $B = 0.25 \, \text{m}^5 \, \text{kg}^{-1} \, \text{s}^{-2}$ in line with swollen rubber networks and f_1 is a correction for the probability of closed loops being formed, which can be written

$$f_1 = \left(1 + \frac{N_{Av}M_n^{1/2}c}{2b^{3/2}N_i[1 - 1/N_i]}\right) \tag{2.59}$$

where b is dependent on the characteristic ratio of the polymer chain (6.38), the skeletal bond length, $l_b = 0.187$ nm, and the molecular weight per skeletal bond, $M_b = 69$ Daltons:

$$b = \frac{3M_b}{2\pi C_\infty l_b^2} \tag{2.60}$$

The calculated curves from Equation (2.58) are shown in Figure 2.12 and give a good fit to the data with between three and four hydrophobes per chain.

2.4.6   Non-Interactive Fillers

There are many situations in which polymer networks contain a filler. The particle size is frequently chosen to be in the colloidal size range.

This is partly for ease of processing but frequently fillers are used to produce colour and/or opacity as well as to improve mechanical properties and it is for this reason also that colloidal particles are used because they provide good light scattering at low addition levels. However we will concentrate on changes in mechanical properties.

The term 'non-interactive filler' means that the filler does not play a role in the crosslinking of the network. Even so fillers can have a marked effect on both elastic properties and wear resistance. Filler particles are usually inorganic or organic particles with a high modulus. For example carbon is used in car tyres.

The simplest approach would appear to be to add the compliances of the two phases in proportion to the volume fractions of each phase:

$$J = \frac{1}{G_N}(1 - \varphi) + \frac{1}{G_f}\varphi \qquad (2.61)$$

Here G_f is the modulus of the filler and for many soft gels the right-hand side is dominated by the first term because $G_f \gg G_N$. We could therefore write the modulus as

$$G \approx \frac{G_N}{(1 - \varphi)} \approx G_N[1 + \varphi + O(\varphi^2)] \qquad (2.62)$$

However this fails to take into account the form of the strain field around the particles. If this is taken into account the modulus of a gel filled with a non-interactive filler can be written as[25]

$$G = G_N[1 + 2.5\varphi + O(\varphi^2)] \qquad (2.63)$$

This is the same as the expression for the viscosity of dilute suspensions (see for example Chapter 3). The expression is however limited to low concentrations of particles, *i.e.* for $\varphi < 0.05$; there are others that can be used throughout the range of volume fractions. Landel[26] proposed that the elastic modulus can be described by

$$G = G_N\left(1 - \frac{\varphi}{\varphi_m}\right)^{-2.5\varphi_m} \qquad (2.64)$$

where φ_m is the maximum concentration that the filler can achieve. For monodisperse spherical particles, the dense random packing value of 0.64 would be more appropriate than the hard sphere transition at ~ 0.5 as the motion of filler particles is likely to become considerably restricted by the polymer network because the mesh size will be smaller than most filler particles. The derivation of Equation (2.64) is straightforward and is achieved by taking an 'effective medium approach'. We know from Equation (2.63) that the initial rate of change in the relative elastic modulus is 2.5, *i.e.*

$$\frac{\partial G}{\partial \varphi} = 2.5 G_N \qquad (2.65)$$

Now if we assume that any volume fraction, that any *replacement* of a small volume of the continuous phase by particles will give rise to the *same rate* of increase in modulus, we have

$$\partial G = 2.5 G \partial \varphi \qquad (2.66)$$

But as our increase in concentration means that we are only able to put particles in place of fluid volume, we must write the concentration addition to the available volume and Equation 2.66 becomes

$$\partial G = 2.5 G \frac{\partial \varphi}{1 - \varphi} \qquad (2.67)$$

Integrating this equation to any volume fraction, φ, with the boundary conditions that the elasticity is equal to the network value when there are no particles present, and that when the volume fraction reaches φ_m the elasticity becomes infinite (although formally we should make it equal to that of the packed filler bed), then gives us Equation (2.64). There is an analogous equation describing the viscosity of suspensions of particles and this will be introduced in Chapter 3. When a value of 0.64 is used for the maximum filler concentration, Equation (2.64) becomes

$$G = G_N \left(1 - \frac{\varphi}{0.64}\right)^{-1.6} \qquad (2.68)$$

An example of the effectiveness of this equation is given by an aqueous HEUR gel made up of a polymer with $M_n = 20 \times 10^3$ Daltons at a concentration of $30\,\mathrm{kg\,m^{-3}}$ filled with a poly(styrene) latex with a particle diameter of $0.2\,\mu m$ at $\varphi = 0.2$. The unfilled gel had a network modulus of 0.4 kPa, whilst the modulus of the filled gel was 0.7 kPa. Equation (2.68) predicts a value of 0.728 kPa. The poly(styrene) particles act as a non-interactive filler because the surface is strongly hydrophobic as it consists mainly of benzene rings and adsorbs a monolayer of HEUR via the hydrophobic groups, resulting in a poly(ethylene oxide) coating that does not interact with the HEUR network. This latter point was further demonstrated at lower filler levels by depletion flocculation of the filler, an event that would not occur with an interactive filler as it requires a tendency for the system to phase separate.

2.4.7 Interactive Fillers

By interactive in this context we mean that the filler can adsorb network polymer and thereby become involved in the crosslinking process, creating a larger number of network links. In some cases the polymer–particle interaction may be the main crosslinking mechanism.

To illustrate how the effect of the adsorption on the modulus of the filled gel may be modelled we consider the interaction of the same HEUR polymer as described above but in this case filled with poly(ethylmethacrylate) latex particles. In this case the particle surface is not so hydrophobic but adsorption of the poly(ethylene oxide) backbone is possible. Note that if a terminal hydrophobe of a chain is detached from a micellar cluster and is adsorbed onto the surface, there is *no net change* in the number of network links and hence the only change in modulus would be due to the volume fraction of the filler. It is only if the backbone is adsorbed that an increase in the number density of network links is produced. As the particles are relatively large compared to the chain dimensions, each adsorption site leads to one additional link. The situation is shown schematically in Figure 2.13. If the number density of additional network links is N_L, we may now write the relative modulus $G_r = G/G_N$ as

$$G_r = \left(1 + \frac{N_L}{N}\right)\left(1 - \frac{\varphi}{0.64}\right)^{-1.6} \tag{2.69}$$

where

$$N_L = \Gamma \frac{N_A}{M_n} \rho_p \varphi \tag{2.70}$$

Here Γ is the adsorbed amount and ρ_p is the density of the filler. The fit to the data is shown in Figure 2.14.

2.4.8 Summary of Polymeric Systems

The above examples show that we can describe the network modulus of polymer gels by using the concept of entropic springs making up the network. In some cases corrections to the network are required to

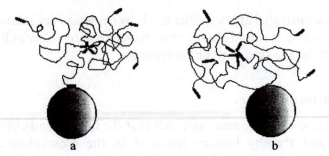

Figure 2.13   (a) *Adsorption of terminal hydrophobic group giving no additional network links;* (b) *adsorption of polymer chain giving an additional network link for each chain adsorbed*

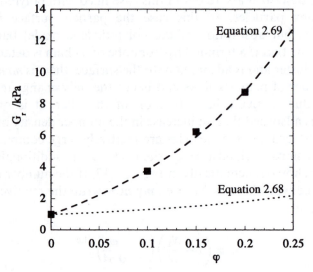

Figure 2.14   *The relative shear modulus, $G_r = G/G_N$, of an HEUR gel filled with polyethylmethacrylate particles. $G_N = 0.4\,kPa$. Experimental points are shown with the curves calculated for a non-interactive and an interactive filler. The amount of adsorbed polymer, $\Gamma = 78\,g/kg$ of filler, gave a good fit to the experimental data*

describe network defects but these can be satisfactorily modelled. Filling a polymer network with either an inert filler or one that interacts with the network by adsorption can also be satisfactorily modelled in a simple fashion. It should be stressed that we are describing the plateau or network modulus, which means that the elasticity is due to the formation of a network and the timescale of the experiments is sufficiently short that no relaxation of the network can occur. The relaxation time is a function of the bond energy so that if the network is formed from covalent bonds the relaxation time is very long. If the bond is weaker, as for hydrophobic bonds for example, shorter relaxation times are found and then it becomes clear experimentally that we have a viscoelastic material and we are looking at a high frequency plateau.

2.5 COLLOIDAL GELS

Dilute colloidal dispersions are fluid systems and if the concentration is increased the mean interparticle separation will eventually approach the range of the interactions between the particles. When this point is reached, a structure is formed which extends throughout the dispersion. Particle motion becomes restricted and viscoelasticity is observed as the relaxation processes become slow. As the particle number density is increased further, the structures become denser and the elastic modulus increases. The details of the particle–particle interactions is a key factor and will be discussed in outline only here. There are other texts which give a detailed account and the reader is referred to these for a full discussion.[12,27,28] The elastic modulus of the colloidal gel is the sum over the structure of all the contributions from the 'colloidal springs'. Recalling that the modulus is the rate of change of force with distance, and that the force is the rate at which the potential energy changes with interparticle separation, the elastic modulus of a 'colloidal spring' is reflected in the curvature of the pair potential curve. Detailed modelling for different types of interaction is given in Chapter 5, and here we will just examine the types of interactions that occur.

2.5.1 Interactions Between Colloidal Particles

The pair potential of colloidal particles, *i.e.* the potential energy of interaction between a pair of colloidal particles as a function of separation distance, is calculated from the linear superposition of the individual energy curves. When this was done using the attractive potential calculated from London dispersion forces, V_a, and electrostatic repulsion, V_e, the theory was called the DLVO Theory (from Derjaguin, Landau, Verwey and Overbeek). Here we will use the term to include other potentials, such as those arising from depletion interactions, V_d, and steric repulsion, V_s, and so we may write the total potential energy of interaction as

$$V_T = V_a + V_d + V_e + V_s \qquad (2.71)$$

It is the convention to denote attractive potentials as negative and repulsive ones as positive.

2.5.2 London–van der Waals' Interactions

London dispersion forces occur between any pair of atoms and are due to the coupling of the oscillations of the electron clouds. This is the same van der Waals' force that acts between any two atoms and is responsible for the non-ideality of inert gases for example. As this is a non-directional attractive force (unlike permanent dipolar interactions for example), the oscillations of large arrays of atoms also couple. There are two approaches used to calculate the interaction energy for colloidal particles.[29] The first is pairwise additivity of the interaction between atoms interacting at a high frequency close to an ionisation condition – the Hamaker approach. This does not take into account the interactions with the atoms surrounding the pair. The problem of multibody interactions can be avoided if the interacting particles are treated as two dielectric masses and the full frequency range is considered – the Lifshitz approach. The result can be expressed in terms of the surface–surface separation of the particle surfaces, h, the particle dimensions, and a coefficient characteristic of the material i interacting across material j, A_{ij}, which is termed the Hamaker constant. The result for a variety of geometries is:

for thick slabs length L and breadth B

$$V_a = -\frac{A_{ij}}{12\pi h^2} LB \tag{2.72a}$$

for spheres of radii R_1 and R_2

$$V_a = -\frac{A_{ij}}{6h} \frac{R_1 R_2}{R_1 + R_2} \tag{2.72b}$$

for a sphere and a plate

$$V_a = -\frac{A_{ij}}{6h} R \tag{2.72c}$$

for crossed cylinders of radii R_1 and R_2

$$V_a = -\frac{A_{ij}}{6h} \sqrt{R_1 R_2} \tag{2.72d}$$

for parallel cylinders of length L

$$V_a = -\frac{A_{ij}}{12h^{3/2}\sqrt{2}} L \left(\frac{R_1 R_2}{R_1 + R_2}\right)^{1/2} \qquad (2.72e)$$

It is perhaps useful to point out at this stage that these interactions are much longer range than those found for two atoms because of the additive nature of the dispersion forces. However the repulsive potential expressed in the Lennard-Jones–Devonshire equation as inversely proportional to the twelfth power of distance is still present at this very short range and will limit the attraction at very short range as the electronic orbital overlap of the surface atoms becomes significant. On this scale the detailed atomic structure of particle surfaces becomes critical and this includes, of course, adsorbed layers of ions, water or other molecules. Although there are some theoretical developments which take some of the detail of surface structure into account, it is usually assumed that we are dealing with smooth surfaces and continuum models will apply. For many situations this is adequate but detailed rheological modelling of systems in which particles are very close should be approached with caution if continuum modelling is being used for the pair potential.

2.5.3 Depletion Interactions

The depletion potential, V_d, occurs when bimodal systems become moderately concentrated. To see a significant effect there needs to be a large size difference and most studies have been done with polymer/ particle systems. It is a prerequisite that the solution polymer does not bind to the surface as is often the case when high levels of surfactants have been used to disperse and stabilise the colloidal system. The origin of the effect is illustrated schematically in Figure 2.15 and is as follows. In a well mixed system we may expect the particles and polymer coils to be randomly distributed throughout the volume. The concentration of polymer in the continuous phase must be close to c^* or higher for a measurable effect as the polymer molecules begin to exert a significant osmotic pressure as they become space filling. Once the particulate phase reaches a concentration at which the average separation is of a similar size to the dimensions of the polymer coil, *i.e.*

$$\bar{h} \le 2R_g$$

then the volume between the particles becomes unavailable to the polymer. The result is that there will be a free energy gain if the particles

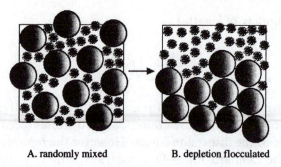

A. randomly mixed   B. depletion flocculated

Figure 2.15 *Equal volumes of a polymer-thickened dispersion. When the particles are aggregated there is a larger solvent volume available to the polymer molecules*

are clustered together into a separate phase giving a balance between the osmotic pressure of the particulate phase and the now more dilute polymer-rich phase. This is clearly a cooperative effect and so the depletion potential should be formally considered from the standpoint of a potential of mean force. However the treatments in the literature normally model the potential in terms of a pair interaction and then assume pairwise additivity. The route[30,31] used for the calculation is to determine the force from the osmotic pressure of the polymer solution multiplied by the cross-sectional area of the volume excluded to the polymer chains between an approaching pair of particles. This is then integrated to give the pair potential. The result is simply the osmotic pressure multiplied by a volume term, which is the volume excluded to the polymer. We may write, for example[31]

$$V_d = -\Pi \frac{4\pi}{3}(a + R_g)^3 \left(1 - \frac{3r}{4(a + R_g)} + \frac{r^3}{16(a + R_g)^3}\right) \qquad (2.73)$$

where Π is the osmotic pressure of the polymer solution and $r = (2a + h)$.

2.5.4 Electrostatic Repulsion

The surfaces of colloidal particles are often charged and these changes can arise from a number of sources. Chemically bound ionogenic species may be found on the surface of particles such as rubber or paint latex particles. Charged species may be physically adsorbed if ionic surface active materials, for example, have been added. A charged surface may occur on a crystal lattice. An example is the isomorphous substitution of lower valency cations such as aluminium for silicon in the lattice structure of clays. A further example is the adsorption of lattice ions which are in excess in solution such as silver iodide which is negatively charged if there is an excess of iodide and is positively charged if silver is in excess. The charge on the particle surface is balanced by a solution charge, which makes up the other half of the *electrical double layer*.

Unlike charges attract and like charges repel each other, so there is a high concentration of counterions attracted to the particle surface whilst co-ions (those with the same sign charge as that of the surface) are repelled. Thermal motion, *i.e.* diffusion, opposes this local concentration gradient so that the counterions are in a diffuse cloud around the particle. Of course particles which have a like charge will also repel each other but the interaction of the particle surfaces will be screened by the counterion clouds between the particles. The interaction potential is a function of the surface potential, ψ_0, and the permittivity of the fluid phase, $\varepsilon = \varepsilon_r \varepsilon_0$, where ε_r is the relative permittivity.[12,27]

For identical spheres at constant surface potential:

$$V_r = \frac{4\pi\varepsilon a^2 \psi_0^2}{r} \exp(-\kappa r) \qquad \text{for } \kappa a < 5$$

$$V_r = 2\pi\varepsilon a \psi_0^2 \{\ln[1 + \exp(-\kappa r)]\} \quad \text{for } \kappa a > 5$$

(2.74a)

The Debye–Hückel decay parameter, κ, which has units of m^{-1}, is a function of the ionic strength, I, and the permittivity:

$$\kappa = \left(\sqrt{\frac{\varepsilon k_B T}{2e^2 N_A I}} \right)^{-1}$$

(2.74b)

In practice we use the electrokinetic potential, ζ, in place of the surface potential as it is readily measurable and will reflect the changes to the surface as a result of adsorbed ionic species.

For two plates:

$$V_r = LB \left(64 n_0 k_B T \tanh^2 \left[\frac{ze\psi}{4k_B T} \right]^2 / \kappa \right) \exp(-\kappa h)$$

(2.74c)

where n_0 is the ion number density in the bulk electrolyte. The expression for a sphere and a plate is derived by multiplying Equation (2.74a) by a factor of 2 and there are expressions in the literature for dissimilar spheres,[32] cylinders[33] and a cylinder and a plate.[34] The calculations can also be carried out using a surface charge that remains constant whilst the potential changes as the surfaces approach.[12,27] The expressions given are the simplest because they are based on the additivity of

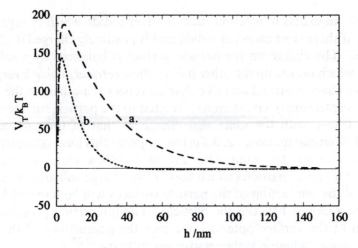

Figure 2.16 *The pair potential calculated for rutile particles with a radius of 100 nm, background electrolyte concentration of 10^{-4} mol dm^{-3} and a ζ-potential of -50 mV. Curve a was calculated for an isolated pair of particles and curve b corresponds to the potential for a pair of particles in a dispersion at a volume fraction of 0.45. Note how the increased electrolyte content due to the counterions introduced with the particles shorten the range of the repulsion enough for a small secondary minimum to be found at h $\sim$ 70 nm*

potentials, and hence ion concentrations, between particles. This means that we are thinking of 'weak overlap'. In many cases of rheological interest this will be a satisfactory approximation whether we are dealing with charge stabilised, weakly flocculated or coagulated colloidal dispersions.

Another important practical problem is encountered when it is recognised that the concept of a 'bulk electrolyte ionic strength' becomes difficult to sustain when a colloid becomes concentrated. The problem is that the counterions that surround each charged particle are in the form of a diffuse cloud and, as the concentration of particles is increased, the ions around neighbouring particles can become involved in the interaction between a particle pair. Thus in a concentrated system the concentration of counterions needs to be added to the background electrolyte content when calculating[12] the value of the Debye–Hückel parameter κ. Figure 2.16 illustrates how the screening of the potential increases with volume fraction.

2.5.5 Steric Repulsion

Steric repulsion is the term used to describe the repulsion between surfaces covered by layers of polymer or non-ionic surfactant. The concept is that as the coated surfaces approach each other a large local osmotic pressure develops as the layers interact.[35] Initially the interaction is a mixing one as the outer layers start to interpenetrate but, as the surfaces approach closer, the interaction includes changes due to restriction of chain conformation as the space is reduced. The interaction is governed by:

1. the number of chains per unit area on the surface, $n_p = \Gamma N_A / M_n A$ (where A is the specific surface area)
2. the thickness of the layer, δ
3. the solvent quality, *i.e.* the value of $(0.5 - \chi)$
4. the strength of the 'anchor' of the chains.

The last point is of critical importance and is a prerequisite for effective stabilisation. The most common means of providing this is to use block copolymers so that the dispersion medium is a poor solvent for part of the chain but a good solvent for the remainder. The anchoring sections commonly vary between 5% and 50% of the chain, depending on the nature of the polymer. Chemical grafting of the stabilising moieties is even more effective but is rarely done in commercial products because of cost. With a strong anchor group, steric stabilisation means that the system is thermodynamically stable and not kinetically stable as with electrostatically stabilised systems. Steric stabilisation is also less sensitive to electrolyte levels and can also be useful in low polarity media. Isrealachvilli[29] gives the interactions between planar surfaces per unit area as:

for low coverage, $\chi = 0.5$ and $\delta = R_g$ (Dolan and Edwards[36])

$$V_s \approx \frac{36}{A} \frac{\Gamma}{M_n} RT \exp\left(-\frac{h}{\delta}\right) \qquad (2.75a)$$

for high coverage, $\chi < 0.5$ and $\delta = n_p^{0.5} (R_g c_\infty)^{0.6}$ (de Gennes[37])

$$V_s \approx \frac{100\delta}{\pi s} \frac{\Gamma}{M_n A} RT \exp\left(-\frac{\pi h}{\delta}\right) \qquad (2.75b)$$

where s is the spacing between anchor points of attached chains and is given by

$$s = (n_p)^{-0.5} = \left(\frac{\Gamma N_A}{A M_n}\right)^{-0.5} \qquad (2.75c)$$

For spherical particles of radii a_1 and a_2 we could use the Derjaguin approximation (see for example reference 29) to calculate the potential:

$$V_s \approx \left(\frac{a_1 a_2}{a_1 + a_2}\right) \frac{200\delta^2}{\pi s} \frac{\Gamma}{M_n A} RT \exp\left(-\frac{\pi h}{\delta}\right) \qquad (2.75d)$$

There are other approximations in the literature that may be more appropriate for particular conditions.[12,27,29,35] In many colloidal systems the value of δ is $2\,\text{nm} < \delta < 10\,\text{nm}$ and until $h < 2\delta$ the pair potential will be dominated by the other types of interparticle interactions, *i.e.* it is of short range, albeit very steep.

It is important to recognise that many of the models that have been tested by experiment have used special systems such as terminally grafted polymer chains. In practical systems using a good steric stabiliser, the osmotic pressure difference of the initial particle interaction will be more a function of the polydispersity in the chains of the stabilising moieties than the particular details of any of the models. Also the surface density of chains is as high as intra-chain interactions will allow, so that the osmotic pressure increases very rapidly with only a small overlap. Hence using the value of δ as a limiting distance of approach is often a satisfactory approximation.

2.5.6   Electrosteric Interactions

Polyelectrolytes are sometimes used as stabilising species and many systems of biological origin have proteins as stabilising species. The result is a strong steric contribution but in addition there is also an electrostatic contribution to the pair potential. The electrostatic term starts from the outside of the layer as the charged species are strongly solvated. Of course, they also repel each other and so an expanded conformation of the layer occurs. As electrolyte is added, or the pH is reduced with -COO^- groups, the repulsion is reduced and the layer contracts. Repulsion still limits the conformation, however, and the potential at the outer surface (the ζ or electrokinetic potential) is very little changed (unlike most ionogenic surfaces) until nearly all of the charge groups are in the undissociated state. This type of stabiliser is a very robust system and ionic surfactant systems also have a component of this type.

The robustness occurs due to the charge being displaced from the particle surface by about the length of the chain in a densely packed layer. Although the displacement is only around a nanometre or two, this can make a very large difference to the shape of the pair potential around

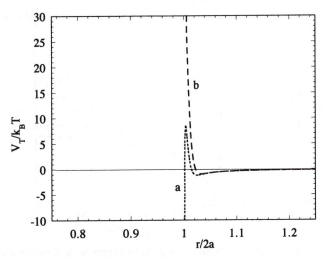

Figure 2.17 *Pair potentials calculated for 50 nm polystyrene particles in 0.1 mol dm^{-3} electrolyte and with a ζ-potential of −30 mV. Curve a is the result for a simple polystyrene surface and curve b was calculated from the model of a 1 nm surfactant layer so that the ζ-potential is taken as occurring at the outer edge of the adsorbed layer. A maximum in the potential of ∼8k$_B$T is insufficient to provide long-term stability and the curves clearly shows how electrosteric stabilisation can achieve this*

the maximum and, because this is close to where the maximum in the force occurs (*i.e.* at the point of inflection), the interparticle forces can be changed significantly. Figure 2.17 shows how this displacement of the origin of the electrostatic interaction combined with a steric contribution can turn a potentially unstable dispersion into a stable one. Another example of the combination of electrostatics and steric barriers is shown in Figure 2.18, which shows the calculated pair potential for large particles of 1 μm in diameter in a sodium chloride solution close to that found in sea water. Under these electrolyte conditions the electrostatics are very short range, starting from a low potential (the measured value was − 12 mV), and the systems would irreversibly coagulate. However the addition of a non-ionic surfactant at a level that would ensure monolayer coverage limits the depth of the attractive well to ∼ 12k$_B$T. This is sufficiently shallow for aggregates of such large particles to be readily broken up by shear forces and the systems are therefore only weakly flocculated.

The data in Figures 2.17 and 2.18 are displayed in terms of the dimensionless centre-to-centre separation of particles, *i.e.* $r/2a = (2a + h)/2a$. This has been done to illustrate another important point: the range of linear elastic response. In a concentrated system, which is showing solid-like or elastic responses, the structure has to be able

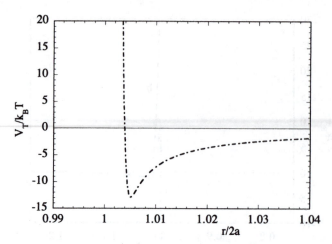

Figure 2.18 *The pair potential calculated for polystyrene particles of radius 500 nm, with a measured ζ-potential of −12 mV in 0.5 mol dm⁻³ electrolyte. A steric barrier of δ = 3.5 nm was used as the particles had monolayer coverage of a monodisperse non-ionic surfactant. This was $C_{12}EO_6$ which represents a dodecyl hydrophobic moiety linked to hexaethylene glycol via an ether link*

to span the volume. This means that the mean value of r is often similar to the particle diameter (or is at least the same order of magnitude). It is therefore useful to think of the strain as a deformation occurring over the distance $2a$ for these 'colloidal springs'. As the energy is the force (stress) times the distance and the spring constant is the rate at which the force changes with distance, the spring constant or modulus is the *curvature* of the pair potential. For this spring modulus to be constant, *i.e.* a linear force–distance curve, the curvature must be constant. (This is the same problem as anharmonicity becoming important in infra-red spectroscopy.) A quick visual inspection of Figures 2.16 and 2.18 shows that for the electrostatically stabilised colloid (curve b in Figure 2.16) the curvature is similar over ∼2% of $r/2a$ whilst for the weakly flocculated system it is less than ∼0.1% of $r/2a$. Bearing in mind that this is only meant as an illustration, it is useful to remember that the critical strain at which non-linearity is observed is an indication of the type of network structure. In practice many electrostatically stabilised systems are linear up to 3 or 4% strain, in contrast to polymer systems which may respond linearly for strains in excess of 50%. It is also the case that the upper limit for weakly flocculated systems is ∼0.1% strain and that for coagulated systems it is difficult to find a linear region at all. This latter condition is because the long-range attraction is eventually limited by very short-range repulsion (interaction of adsorbed species occurring at ∼h^{-12} as in atomic interactions. This results in very sharp curvature indeed.

2.6 REFERENCES

1. C.W. Macosko, *Rheology: Principles, Measurements and Applications*, Wiley-VCH, New York, 1994.
2. W.G. Hoover and F.H. Ree, *J. Chem. Phys.* 1968, **49**, 3609.
3. J.G. Kirkwood, *J. Chem Phys.* 1939, **7**, 919; B.J. Alder and T.E. Wainwright, *Phys. Rev.* 1962, **127**, 359.
4. See for example, P.W.Atkins, *Physical Chemistry*, 4th edn, Oxford University Press, Oxford, 1990.
5. C. Croxton, *An Introduction to Liquid State Physics*, Wiley, London, 1975.
6. D. Tabor, *Gases, Liquids and Solids and Other States of Matter*, 3rd edn, Cambridge University Press, Cambridge, 1991.
7. N.W. Tschoegl, *The Phenomenological Theory of Linear Viscoelastic Behaviour*, Springer-Verlag, Berlin, 1989.
8. M. Doi and S.F. Edwards, *The Theory of Polymer Dynamics*, Oxford University Press, Oxford, 1986.
9. P.J. Flory, *Statistical Mechanics of Chain Molecules*, Interscience, New York, 1969.
10. H. Yamakawa, *Modern Theory of Polymer Solutions*, Harper and Row, New York, 1971.
11. P.J. Flory, *Principles of Polymer Chemistry*, Cornell University Press, Ithaca, NY, 1953.
12. W.B. Russel, D.A. Saville and W.R. Schowalter, *Colloidal Dispersions*, Cambridge University Press, Cambridge, 1989.
13. S.H. Hyon, W.I. Cha and Y. Ikada, *Polymer Bull.* 1989, **22**, 119.
14. R.J. Roark and W.C. Young, *Formulas for Stress and Strain*, 5th edn, McGraw-Hill, New York, 1975.
15. G.W. Scherer, *J. Non-Cryst. Solids* 1992, **142**, 18.
16. M.S. Green and A.V. Tobolsky, *J. Chem. Phys.* 1945, **14**, 80.
17. H.M. James and E.J. Guth, *J. Chem. Phys.* 1947, **15**, 669.
18. F. Tanaka and E.J. Edwards, *J. Non-Newtonian Fluid Mech.* 1992, **43**, 247.
19. E.J. Glass, *Polymers in Aqueous Media: Performance Through Association*, Adv. Chem. Series **223**, American Chemical Society, Washington, 1989.
20. S. Lam, Ph.D. thesis, Bristol University, 1994. (We gratefully acknowledge the help rendered by Dr P. Sperry of Rohm and Haas and Prof. W.B. Russel of Princeton University during this work.)
21. T. Annable, R. Buscall, R. Ettalie and D. Whittlestone, *J. Rheol.* 1993, **37**, 695.
22. B. Cabane, R. Dupplesix and T. Zemb, *J. Physique* 1985, **46**, 2161.
23. J.W. Goodwin, in *Industrial Water Soluble Polymers*, ed. C.A. Finch, Special Publication 186, The Royal Society of Chemistry, Cambridge, 1996.
24. J.W. Goodwin, R.W. Hughes, C.K. Lam, J.A. Miles and B.C.H. Warren, in *Polymers in Aqueous Media: Performance Through Association*, ed. E.J. Glass, Adv. Chem. Series **223**, American Chemical Society, Washington, 1989.
25. H.M. Smallwood, *J. Appl. Phys.* 1944, **15**, 758.
26. R.F. Landel, *Trans. Soc. Rheol.* 1958, **2**, 53.
27. R.J. Hunter, *Foundations of Colloid Science*, Vols 1 and 2, Clarendon Press, Oxford, 1989.
28. J. Lyklema, *Fundamentals of Interface and Colloid Science*, Vols 1 and 2, Academic Press, London, 1991.
29. J. Israelachvili, *Intermolecular and Surface Forces*, Academic Press, London, 1985.
30. S. Asakura and F. Oosawa, *J. Chem. Phys.* 1954, **22**, 1255.
31. A. Vrij, *Pure Appl. Chem.* 1975, **48**, 471.
32. R. Hogg, T.W. Healy and D.W. Fursteneau, *Trans. Faraday Soc.* 1966, **62**, 1638.
33. M.J. Spaarnay, *Recueil* 1959, **78**, 680.
34. I. Callaghan, PhD thesis, University of Bristol, 1973.

35. D.H. Napper, *Polymeric Stabilisation of Colloidal Dispersions*, Academic Press, London, 1983.
36. A.K. Dolan and S.F. Edwards, *Proc. R. Soc. London* 1974, **A337**, 509.
37. P.G. de Gennes, *Adv. Coll. Interface Sci.* 1987, **27**, 189.

CHAPTER 3

Viscosity: Low Deborah Number Measurements

3.1 INITIAL CONSIDERATIONS

Flow of materials occurs when the deformation that we have produced by applying a stress to the material is not recovered fully when the stress is removed. This is a familiar concept when we think of the behaviour of such common materials as light oils or water. Indeed it would be surprising if any of the deformation was recovered, although we shall see that with some solutions and dispersions this is not the case.

We have seen that with high Deborah number experiments the work done when a stress was applied was stored throughout the microstructure as the components were moved into a higher energy configuration. At Deborah numbers $\ll 1$, the thermal or diffusive motion is sufficient for the microstructure to remain at, or very close to, the low energy configuration. In other words, the work done is continually dissipated. This gives us the mechanical equivalence of heat and results in the concept of an internal friction coefficient. When the stress that we apply is kept constant, the deformation will occur at a constant rate. As an example of this, we have a sample of our material in a cup with a bob immersed in it, as illustrated in Figure 3.1. When we apply a constant torque to the bob, it will rotate at a constant angular velocity and so the strain will be increasing at a constant rate. The tangential velocity of the bob surface is $v = \mathrm{d}x/\mathrm{d}t$, and so the rate of strain, $\mathrm{d}\gamma/\mathrm{d}t = \dot{\gamma}$, is $\mathrm{d}/\mathrm{d}t(\mathrm{d}x/\mathrm{d}y) = \mathrm{d}v/\mathrm{d}y$ or the velocity gradient. A typical rheogram for water at $25\,°\mathrm{C}$ is shown in Figure 3.2. This rheogram shows a linear relationship between the shear stress and the shear rate and a fluid that exhibits this type of behaviour is known as a Newtonian fluid. The slope of the line gives the coefficient of (shear) viscosity, $i.e.$ $\eta = \frac{\sigma}{\dot{\gamma}}$, which in this case is $\sim 9 \times 10^{-4}\,\mathrm{Pa\,s}$. The reciprocal of the viscosity is known as

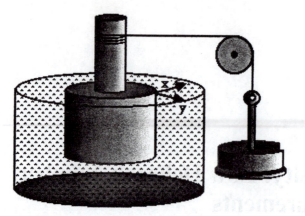

Figure 3.1   *Simplest cup and bob geometry. The x-direction is tangential to the moving surface and y is perpendicular*

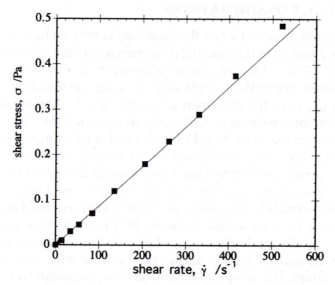

Figure 3.2   *Rheogram for water at 25 °C. The slope gives a value for the viscosity of 8.85×10^{-4} Pa s*

the fluidity, ϕ, of the material. This term is less frequently used than viscosity although there may be occasions when it is useful to think in these terms.

It is important to note that the reciprocal of the shear rate is the time taken for unit strain to occur in the material and is the characteristic timescale for our experiment. As long as the microstructure can reorganise by thermal motion in a shorter timescale, the value of *De* is less than 1 and the structure will remain in a configuration that is close to its equilibrium structure. The energy of course is continually dissipated and we can calculate the rate from the force and the distance moved per second:

$$\dot{E} = \sigma\dot{\gamma} \qquad (3.1)$$

The constitutive equation or equation of state for our Newtonian fluid is

$$\sigma = \eta\dot{\gamma} \qquad (3.2)$$

and so the rate of energy dissipation for a Newtonian material is now

$$\dot{E} = \eta\dot{\gamma}^2 \qquad (3.3)$$

We can take as an example a Newtonian oil, hexadecane say, which has a viscosity of 3.34×10^{-3} Pa s at 20 °C, in our rheometer with a sample size of 2 g of oil. We are going to shear this sample at $300\,s^{-1}$ for 100 s. The energy input is 3340 J. The heat capacity of the oil is $\sim 500\,J\,mol^{-1}\,K$ and so the temperature increase would be 6.7 °C if our rheometer was a very poor heat conductor (*i.e.* an adiabatic system). The viscosity decreases exponentially with temperature and so clearly there is a requirement for good thermostatting so that the experiment is isothermal. But even if the rheometer is a good heat conductor, large samples of viscous organic materials may not remain at a uniform temperature across the sample!

The energy changes that occur when a fluid flows are due to the sum of

1. changes in potential energy (*e.g.* changes in height)
2. changes in pressure (reversible pressure volume term)
3. changes in kinetic energy (from the momentum changes)
4. the viscous or 'frictional' losses

Formally this is expressed by the Bernoulli equation (see for example reference 1), but in the laboratory we have the ability to work under conditions such that we can minimise some of these. For example, in a small sample in our simple viscometer pictured in Figure 3.1, a small volume element of fluid would not be expected to change in height in the cup. We usually regard liquids as incompressible but we should bear in mind that this is not strictly the case. The faster a fluid flows, the lower is the pressure. The best example of this is the lift generated by an aerofoil moving through a fluid. This will also not raise any problems in our simple viscometric flow so let us focus for a moment on points 3 and 4 above for a flowing liquid. The amount of energy dissipated in one second due to the flow is the force multiplied by the velocity. The force is given by the momentum of the fluid element, which is simply the mass times the velocity. The mass of fluid that we are considering is calculated from the density, r, and the volume of fluid passing through the element. Hence we can say that the force due to the momentum is

$$F_m \propto mv$$
$$\text{or } F_m \propto \rho a^2 v^2 \tag{3.4}$$

The velocity times the Stoke's friction factor gives the viscous resistive force:

$$F_v \propto v\eta a \tag{3.5}$$

It is interesting to ask what we observe when one or the other of these forces dominate. The ratio of these two forces gives us the dimensionless group known as the Reynolds number, Re:

$$Re = \frac{\rho a v}{\eta} \tag{3.6}$$

We find experimentally that when $Re \geq O(10^3)$ the flow is no longer laminar, $i.e.$ the flow becomes unstable and vortices are formed. These are first seen at the boundaries between the fluid and solid surfaces. These chaotic flows should be avoided in our laboratory equipment in the course of viscometric characterisation. The conclusion that we draw is that we should make measurements at low Reynolds numbers in order to ensure that only the viscous dissipation is making a significant contribution to our measurements.

3.2 VISCOMETRIC MEASUREMENT

A wide variety of viscometers suitable for liquids are currently available, often with computer control. Many quality control laboratories use simple, cheap, robust instrumentation, which performs quite adequately in a day to day context. However these instruments can have a very narrow range and do not always give well-defined shear rates. This makes them less suitable for research and development work and we will not consider them further here. Figure 3.3 shows schematically the two main instrument types in common use: controlled stress, where the stress is applied electrically via a motor leaving us to measure the strain; and

Controlled stress   Controlled strain

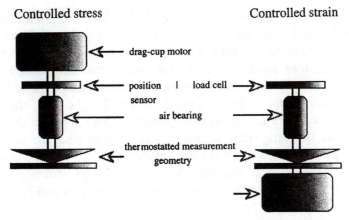

Figure 3.3 *Schematic of currently available rheometers. Computer control of stress, strain and temperature are standard*

the controlled strain instruments where a strain is imposed and the stress is computed from the deformation of a calibrated spring system.

Both types of instrument commonly have an air bearing to give greater sensitivity with low viscosity samples, although the viscosity of water or simple organic liquids is usually at the lower end of the range. Most instruments would benefit from improvements in thermostatic control.

A range of different measuring geometries is available in addition to simple parallel plates. The designs are such that the shear rate is approximately constant throughout the sample. Commonly used geometries are shown in cross-section in Figure 3.4.

3.2.1 The Cone and Plate

A flat plate forms the lower element and the upper element is conical in shape with a cone angle close to π rad. Advantages of such an arrangement are that only small sample volumes are needed, the mass of the cone can be kept low to minimise the moment of inertia, and they are easy to

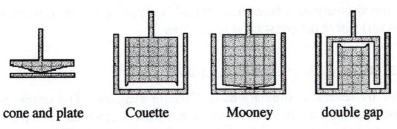

cone and plate   Couette   Mooney   double gap

Figure 3.4 *Vertical cross-sections of common measuring geometries used to provide simple viscometric flow*

clean. The point of the cone is truncated to reduce friction and the gap between the two elements is adjusted so that the point of the cone just contacts the plate. In addition this small truncated region makes the gap setting procedure easier to carry out. Any error introduced by the truncation is negligible. More important is the included angle between the two elements, which should normally be in the range $0.005 < \alpha < 0.002$ rad. The velocity gradient at any point on the plate is found from the tangential velocity at that point and the separation between the surfaces. For an angular velocity of Ω rad s^{-1}, the tangential velocity is Ωr. If the included angle is α, then the gap at r is $z = r \tan \alpha \approx r\alpha$ if the angle is small. The shear rate in a *small* angle cone and plate is then

$$\dot{\gamma}_{cp} \approx \frac{\Omega}{\alpha} \tag{3.7}$$

What we measure experimentally at a given shear rate is the moment of the force on the cone, which gives us the twist in the spring or load cell. That is, the sum of the forces on each element dr wide multiplied the distance from the centre r:

$$M_{cp} = \int_0^a \sigma 2\pi r^2 \mathrm{d}r \tag{3.8}$$

that is

$$M_{cp} = \frac{2\pi a^3}{3}\sigma \tag{3.9}$$

If D is the angular displacement of the spring with a spring constant of C_s, *i.e.* the torque required to give unit angular displacement, then the stress is

$$\sigma = \Delta\left(\frac{3C_s}{2\pi a^3}\right) \tag{3.10}$$

and the viscosity of a Newtonian liquid is simply $\eta = \Delta/\Omega$ (instrument constant) where the constant is $(3C_s\alpha/2\pi a^3)$.

3.2.2 The Couette or Concentric Cylinder

There are several combinations of Couette available. The simple bob shown in Figure 3.4 has a concave base, which is designed to trap air in order to reduce the drag contribution from the base. This geometry usually has a larger area and delivers a larger torque for a given sample condition than a cone and plate. The sample size is larger than for a cone and plate and it may be easier to reduce problems of evaporation by coating the surface with a thin layer of low viscosity insoluble liquid. The gap between the two cylinders should be small in order to get as narrow a range of shear rates as possible. The analysis is simple and at this stage we will only consider a Newtonian liquid so that $\eta = \sigma/\dot{\gamma}$.

If the length of the bob is L and the radii of the cylinders are R_i and R_o for the inner and outer cylinders respectively, the local change in tangential velocity across a small liquid element dr at r from the centre line of the bob is the local shear rate $= r d\omega/dr$ and the shear stress in the element is $\sigma = \eta r d\omega/dr$ for our Newtonian liquid. The moment of the drag force is the sum of the stress in each element multiplied by the area of the cylindrical surface times the distance r, *i.e.*

$$M_{cc} = \text{sum of } (2\pi L r^2 \text{ stress from each element}) \qquad (3.11)$$

So to calculate the angular velocity of the outer cylinder, we have to sum all the contributions from the static cylinder to the moving one:

$$\int_0^\Omega d\omega = \frac{M_{cc}}{2\pi L \eta} \int_{R_i}^{R_o} \frac{dr}{r^3} \qquad (3.12)$$

so that

$$\Omega = \frac{M_{cc}}{4\pi L \eta} \left[\frac{1}{R_i^2} - \frac{1}{R_o^2} \right] \qquad (3.13)$$

We can rearrange this in terms of the viscosity to give the Margules equation:

$$\eta = \frac{M_{cc}}{4\pi L \Omega} \left[\frac{1}{R_i^2} - \frac{1}{R_o^2} \right] \qquad (3.14)$$

We can now calculate the local shear rate in this Newtonian liquid at any point in the gap. The stress at r is

$$\sigma = \frac{M_{cc}}{2\pi r^2 L} \qquad (3.15)$$

and so the shear rate is

$$\dot{\gamma} = \frac{\sigma}{\eta} = \frac{2\Omega}{r^2} \left[\frac{1}{R_i^2} - \frac{1}{R_o^2} \right]^{-1} \qquad (3.16)$$

This equation enables us to calculate the shear rate at each cylinder surface and the mean value.

When $r = R_i$

$$\dot{\gamma}_i = 2\Omega \frac{R_o^2}{R_o^2 - R_i^2} \qquad (3.17)$$

and at $r = R_o$

$$\dot{\gamma}_o = 2\Omega \frac{R_i^2}{R_o^2 - R_i^2} \qquad (3.18)$$

and the mean value

$$\langle \dot{\gamma} \rangle = \Omega \frac{R_o^2 + R_i^2}{R_o^2 - R_i^2} \qquad (3.19)$$

A quick calculation will demonstrate that if the gap is 5% of the outer cylinder radius, then the shear rate varies by 5% across the gap and so small gaps should be used if possible. If the gap is small, we can make the approximation $(R_o + R_i) \approx 2R_o \approx 2R_i$ and then the shear rate is

$$\dot{\gamma} \approx \frac{\Omega R_o}{(R_o - R_i)} \qquad (3.20)$$

The Mooney arrangement of a bob with a conical base is an attractive design as it is relatively easy to fill and uses the base area to enhance the measurement sensitivity. However the cone angle must be such that the shear rates in both the cone and plate and concentric cylinder sections are the same. This means that the gap between the cylinders must be very slightly larger than the gap at the edge of the cone and plate if a constant shear rate is required. Unfortunately the DIN standard bob is poor in this respect.

The double gap geometry provides a high sensitivity with a light construction. However the engineering is subtle in that the inner gap must be slightly smaller than the outer one if a uniform shear rate is to be found throughout the sample. It is also less satisfactory in terms of the onset of flow instabilities. Taylor[2] showed that the Reynolds number for the onset of such instabilities is lower if the inner cylinder is made to rotate and, with the double gap arrangement, this occurs in the inner section. This will significantly lower the upper range of shear rate available, although not to the level in a cone and plate with a large included angle.

So far we have restricted our discussion to Newtonian liquids, but the analysis will change somewhat if the systems are non-Newtonian. A useful illustration of the problems that arise is the case of a Bingham plastic. This gives us a linear response, as does a Newtonian liquid, but in this case there is an intercept or yield stress. The constitutive equation for a Bingham plastic is

$$\sigma = \eta(\infty)\dot{\gamma} + \sigma_B \qquad (3.21)$$

Here the yield stress is the Bingham yield value and the value of $\eta(\infty)$ is the linear value reached at high shear, often referred to as the plastic viscosity. The calculation of the material behaviour follows the same route as with the Newtonian case so:

moment = sum of moments from all the elements

i.e.

$$M_{cc} = \text{sum of } (2\pi L r^2 \text{ stress from each element}) \qquad (3.22)$$

We must use the constitutive equation for the Bingham plastic for the stress. This then gives the angular velocity of the outer cylinder from:

$$\int_0^\Omega d\omega = \int_{R_i}^{R_o} \left(\frac{M_{cc}}{2\pi L \eta(\infty)r^3} - \frac{\sigma_B}{\eta(\infty)r} \right) dr \qquad (3.23)$$

giving the Riener–Riwlin equation for plastic flow in a Couette:

$$\Omega = \frac{M_{cc}}{4\pi L \eta(\infty)} \left[\frac{1}{R_i^2} - \frac{1}{R_o^2} \right] - \frac{\sigma_B}{\eta(\infty)} \ln\left(\frac{R_o}{R_i} \right) \qquad (3.24)$$

Clearly the Riener–Riwlin equation reduces to the Margules equation when the Bingham yield value is zero, but there is an important consequence in that it is assumed that *all* the material is flowing, *i.e.* the shear stress at the wall of the outer cylinder must be

$$\frac{M_{cc}}{2\pi L R_o^2} \geq \sigma_B \qquad (3.25)$$

If this is not the case, a region near the outer wall will not flow and only part of the sample will be sheared and so the actual shear rate will be

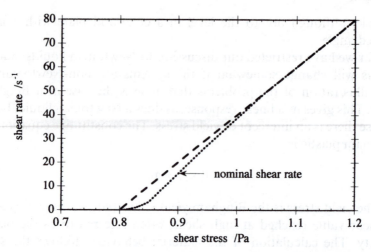

Figure 3.5 *Calculated rheograms for a Bingham plastic with $\eta(\infty) = 5 \times 10^{-3}\,Pa\,s$ and $\sigma_B = 0.8\,Pa$. Note how the nominal shear rate is significantly lower than the actual value resulting from a given applied stress*

higher than the nominal value. The point at which the flow commences is readily found from

$$\left(\frac{M_{cc}}{2\pi L\sigma_B}\right)^{1/2} = r \qquad (3.26)$$

Figure 3.5 depicts the curves that would result from a controlled stress experiment, showing that there is significant curvature if the shear rate is simply calculated from cylinder sizes. This illustrates the importance of using the appropriate equation of state when analysing the viscometer response.

3.3 THE MOLECULAR ORIGINS OF VISCOSITY

Up to this point we have characterised our materials as continua and defined the material parameters. This may be all that is required for engineering purposes or quality control needs. Whenever a modification of the behaviour is sought, a deeper understanding of the origins of the response is required. It was pointed out in Chapter 1 that the rheology is controlled by the atomic or molecular interactions in the system, and this brings the subject properly into focus for the chemist.

3.3.1 The Flow of Gases

Let us consider what happens when a stress is applied to a surface, as illustrated in Figure 3.4, and the fluid between the surfaces is a gas. Prior to the stress being applied, the gas molecules are moving randomly with thermal motion at a velocity described by the kinetic theory of gases.[3] When the stress is applied, the inner surface moves and attains a constant velocity and the proportionality between the stress and rate of strain gives the gas viscosity from $\eta = \sigma/\dot{\gamma}$. The motion of the cylinder sets up the velocity gradient in the gas and, if we visualise what is occurring at any cylindrical surface between, and parallel to, the surfaces, the origin of the resistance to motion becomes clear. The average momentum of the molecules in a plane is the product of their mass and the velocity at the plane. The flux of molecules *through* the cylindrical surface from diffusion means that there is also a flux of momentum which will be transferred to the moving surface. This will always result in a decrease in momentum as the molecules are coming from layers that are moving more slowly (the molecules colliding with the static surface will lose their momentum). Hence a force (stress) must be maintained to keep the motion constant. A clear and simple derivation is given by Tabor[4] from the kinetic theory of gases to give the viscosity:

$$\eta = \frac{1}{3}\frac{m\bar{c}}{\pi\sigma_e^2} \tag{3.27}$$

Here m is the mass of a molecule with an effective diameter of σ_e and a mean speed of $\bar{c}$ (which of course is also a function of mass). This equation was first derived by Maxwell in 1860. We can immediately see the importance of the nature of the molecules through their mass and through the effective diameter, which is not just a geometric size but includes a contribution from the interaction with its neighbours.

3.3.2 The Flow of Liquids

The liquid state is a condensed state, so each molecule is always interacting with a group of neighbours although diffusing quite rapidly. As a result, although momentum through a shear plane still occurs, it is a small contribution when compared to the 'frictional resistance' of the molecules in adjacent layers. It is the nature of this 'frictional resistance' that we must now address and it will become clear that it arises from the intermolecular forces. The theories of the viscosity of liquids are still in an unfinished state but the physical ideas have been laid down. The first point to remember is that the intermolecular forces are strong. For example, the surface tension of a pure liquid (γ_s) is simply the work done in creating unit surface area. So if we were to pull a section of liquid apart we would expend $2\gamma_s$ J m^{-2}. When we scale this to the size of, say, a water molecule (0.25 nm); the tensile stress required is of the order of 600 MPa (equivalent to 6000 atmospheres)!

A consequence of the intermolecular forces is that any molecule in the liquid state has a large coordination number so that the structure has a low energy configuration. In other words there is *short-range* order in the liquid state. A significant amount of work must be done to remove a molecule from this structure – the latent heat of vaporisation.

What happens when we make a liquid flow by, say, applying a shear stress? The molecules move relative to one another if flow occurs, but unlike the gaseous state the molecules are in continuous interaction with their neighbours. This means that the molecules in the structure change their nearest neighbours whilst maintaining the same average coordination number. The rate at which a liquid structure can rearrange is a function of the strength of the intermolecular interactions and the amount of free volume in the system. Experimentally we observe that the viscosity of molecular liquids changes rapidly (exponentially) with temperature and pressure.

3.3.3   Density and Phase Changes

Let us first consider how the density of the condensed phase changes with temperature. As our material in the vapour phase is cooled at constant pressure the density increases until the boiling point is reached. Further cooling then allows us to differentiate between the vapour and liquid states by the formation of a boundary. Further cooling increases the liquid density but at a much slower rate than that of the gaseous phase. The density of many liquids can be described by a simple linear equation over a wide temperature range:[5]

$$\rho_T = \rho_o \left[1 - \frac{(T - T_o)}{A_1} \right] \tag{3.28}$$

Here T_o is a reference temperature and if this is ~ 273 K the coefficient $A_1 \sim 10^3$. The temperature may then be lowered until the melting point, T_m, is reached. At this point a crystalline solid is formed, the density of which will usually be within $\sim 10\%$ of the liquid. The density can be described by a similar equation to Equation (3.28) but the coefficient is quite different, *i.e.* $A_s \sim 10^{15}$! The origin of the thermal expansion in a crystal is simply the increasing oscillations of the molecules about their mean positions. The intermolecular forces are quite strong and of short range and hence the motion is restricted. It is useful to consider further (i) the melting and (ii) the solidification processes.

As the temperature is raised, the vibrational energy increases, because it is $k_B T$ in each direction. If we have a simple cubic crystal in which the intermolecular spacing is r then the molar volume is $N_a r^3$. The Young's modulus for the crystal is Y and we assume a Hooke's law spring. We can define the local stress as the applied force per molecule, F_m, divided by r^2, giving a local strain of x/r where x is the extension caused by the oscillation. Hence:

$$\frac{F_m}{r^2} = Y\frac{x}{r} \quad \text{or} \quad F_m = rYx \qquad (3.29)$$

and the energy is

$$k_B T = \int_{-x/2}^{+x/2} F_m dx \quad \text{or} \quad k_B T = rY \int_{-x/2}^{+x/2} x dx$$

i.e.

$$k_B T = \frac{1}{2} rYx^2 \qquad (3.30)$$

Lindemann[6] said melting would occur when x became some significant fraction, α, of r, say of the order of 0.1. Now Equation (3.30) can be written in terms of the molar volume, V_m, and α:

$$RT_m = \frac{1}{2} YV_m \alpha^2 \qquad (3.31)$$

where R is the gas constant. As $V_m = M/\rho$, the melting point can be written as

$$T_m = \frac{YM}{2\rho} \alpha^2 \qquad (3.32)$$

This very approximate model gives reasonable agreement with such disparate materials as iron, quartz and sodium chloride with $\alpha \sim 0.14$. Alternatively, of course, if we know the melting point, we can obtain an estimate of the elastic modulus.

The crystallisation process, whether from the melt or a concentrated solution, is also an interesting process that has a bearing on the rheological behaviour. On cooling, crystal nuclei will eventually form and then grow to form macroscopic crystals. In some circumstances either or both of these process may not occur readily. Nucleation requires that a few molecules become locally organised to such an extent that the energy is minimised. This may be difficult to achieve if we have large complex molecules such as a high molecular weight polymer. These usually only show crystalline regions in which a portion of any one molecule is involved. However small molecules may also be slow to nucleate. Glycerol and solutions of monosaccharides are common examples. The details of molecular shape and hydrogen bonding can mean that nuclei are slow to form because a lot of cooperative motion is required and, if the system is being cooled at a steady rate, marked supercooling may occur. This will enhance the nucleation rate but the viscosity will be increasing simultaneously and this will reduce molecular motion and inhibit crystal growth.

The granulation of honey provides a good example of these phenomena. The glucose/fructose ratio is important. The glucose is in the pyranose ring structure and crystallises readily whereas the fructose occurs as furanose rings as well as pyranose rings and is very much more difficult to nucleate. As honey is cooled below 25 °C the nucleation increases and granulation becomes rapid. However, the rate reaches a maximum at ~ 14 °C and holding at lower temperatures slows crystallisation as the viscosity rapidly increases. Figure 3.6 shows the viscosity of glycerol as a function of temperature.[7] The viscosity increases steadily

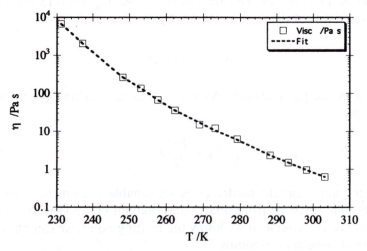

Figure 3.6   *The viscosity of glycerol as a function of temperature.[7] T_m is 278 K and at lower temperatures the glycerol is a supercooled liquid*

as the temperature is decreased well below the melting point. Over most of the range displayed in Figure 3.6, the glycerol is a supercooled liquid.

Further cooling of a supercooled liquid increases its density and viscosity until solid-like behaviour is observed. At this point the material is in a glassy state and the temperature is the glass transition temperature, T_g. The viscosity at this point is now extremely high so that we can say $\eta > 10^{12}$ Pa s for example.[5] This viscosity value is high but not infinite and so we must recognise that *slow* relaxation may occur and so the value that we assign to the T_g will be a function of our patience with the experiment.[8] Figure 3.7 shows a schematic phase diagram of the specific volume as a function of temperature. Note that the equilibrium glass has a denser structure than that formed at the T_g by normal cooling. The density is, however, higher than the crystalline solid as the order is shorter range in the glass. At T_o the transition is second order and the molecular volume is defined as V_o and so at any point in the liquid region we may define the free volume V_f as

$$V_f = \frac{1}{\rho_T} - V \qquad (3.33)$$

and we see that the free volume increases monotonically as we heat our material through the liquid region.

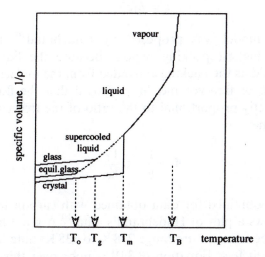

Figure 3.7 *Schematic phase diagram showing the specific volume as a function of temperature (adapted from reference 5)*

3.3.4 Free Volume Model of Liquid Flow

This stems from the idea that when the free volume in the liquid is zero, motion is restricted to local molecular oscillation so flow cannot occur and the material is a solid. As the temperature is increased, the decrease in density means that the free volume is increasing and flow now becomes possible with the viscosity decreasing as the free volume increases. Doolittle[9] suggested an exponential relationship:

$$\eta = A e^{\left(\frac{V_o}{B V_f}\right)} \qquad (3.34)$$

Here A and B are constants and a good fit is achieved over a wide temperature range. Using $\Delta T = T - T_o$ in Equation (3.28), we may write the ratio of the molecular volume to the free volume as

$$\frac{V_o}{V_f} = \frac{A_1}{\Delta T} - 1$$

and then Equation (3.34) can be rewritten as

$$\ln \eta + \ln\left(2.718 \frac{e^B}{A}\right) = \frac{A_1}{\Delta T}$$

or with $B_1 = \dfrac{1}{\left(2.718 \frac{e^B}{A}\right)}$:

$$\eta = B_1 e^{\frac{A_1}{\Delta T}} \qquad (3.35)$$

An alternative model was proposed by Hildebrand[10] in which he considered the highest packing region. Because the liquids became rapidly more fluid as the packing expanded from the molecular volume, i.e. at low levels of free volume, he proposed that the fluidity of the liquids was directly proportional to the ratio of the free volume to the molecular volume:

$$\Phi = \frac{1}{\eta} = C \frac{V_f}{V_0} \qquad (3.36)$$

A good fit was obtained for data obtained with carbon tetrachloride. Figure 3.8a shows a plot of Hildebrand's data.[10] A good linear fit was obtained over the temperature range 278 K to 338 K; note however that the exponential fit [e.g. Equation (3.34)] is poor over this range when plotted as fluidity. A good fit is found when the data is plotted as the

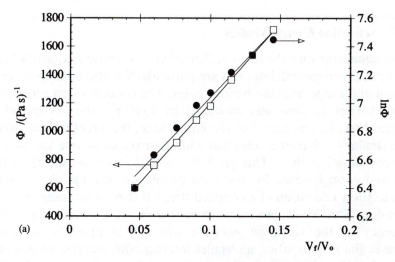

Figure 3.8a *Data for CCl₄ for the fluidity versus the ratio of free to filled volumes (redrawn from Hildebrand[10])*

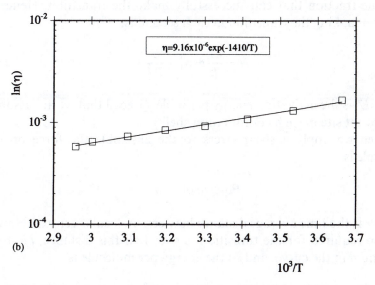

Figure 3.8b *The temperature dependence of the natural logarithm of the viscosity of CCl₄*

natural logarithm of viscosity versus the reciprocal of the temperature in Figure 3.8b indicating a good fit to the form:

$$\eta = B_2 e^{\frac{-A_2}{T}} \tag{3.37}$$

The exponential dependence on temperature was taken by Hildebrand to be due to the variation of the free volume ratio with temperature.

3.3.5 Activation Energy Models

Fits to equations with the form of Equation (3.37) have frequently been used to fit experimental data, and are particularly useful at temperatures at the liquid range near the boiling point. The concept of an activation energy barrier to flow was modelled by Eyring[11] and his model is outlined here. In this model of the liquid state, the structure was taken to be similar to that of a solid but with approximately one vacancy in each coordination shell. This gives the $\sim 10\%$ lower density than the solid and enough space for sufficient movement of molecules to take place to allow relaxation of an applied stress. If there is no energy barrier to motion, a molecule can move from one site to a vacant site at the frequency of the vibration $\sim k_B T/h$ where h is Planck's constant. However the neighbouring molecules interact with our reference molecule and oppose the motion. This results in an activation energy barrier as shown schematically in Figure 3.9. The Boltzmann distribution will give the fraction that can successfully make the transition. Hence the fraction making the transition in each second is:

$$N = \frac{kT}{h} \exp\left(-\frac{E^*}{RT}\right) \tag{3.38}$$

where E^* is the activation energy per mole. (Recall that we are assuming one vacant site in each coordination shell.)

When we apply a shear stress to the ensemble the force on each molecule is

$$\text{Force/mol} = \sigma \bar{r}^2$$

where $\bar{r}$ is the time average distance between molecular centres. Now the motion required for the transition is over half this distance, *i.e.* to the maximum in the curve, and so the energy per molecule is

$$\text{Energy/mol} = \frac{\sigma \bar{r}^3}{2}$$

This energy assists the motion to the right in Figure 3.9 and would oppose it in returning to the left. Using V_m as the minimum volume occupied by Avogadro's number of molecules enables us to write the jump frequency as

$$N = \frac{k_B T}{h} \exp\left(-\frac{E^* \pm (\sigma V_m)/2}{RT}\right) \tag{3.39}$$

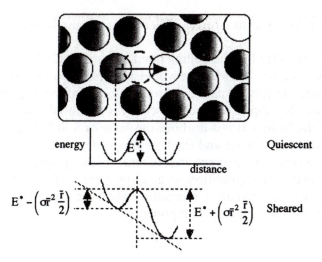

Figure 3.9 *Activated flow*

The shear rate is the distance moved in each second over the distance separating the layers:

$$\dot{\gamma} = \frac{\bar{r}N}{\bar{r}'} \approx N$$

where $\bar{r}'$ is the perpendicular distance between moving layers.
Recalling that $\sinh x = (e^x - e^{-x})/2$, we can write Equation (3.39) as:

$$\dot{\gamma} = \frac{k_B T}{h} \exp\left(-\frac{E^*}{RT}\right) 2\sinh\left(\frac{(\sigma V_m)}{2RT}\right) \tag{3.40}$$

Also recalling that $\sinh x \rightarrow x$ as $x \rightarrow 0$ gives us the zero shear viscosity as

$$\eta(0) = \frac{h}{V_m} \exp\left(\frac{E^*}{RT}\right) \tag{3.41}$$

3.4 SUPERFLUIDS

Helium provides an interesting variant of normal liquid behaviour. When ^{4}He is cooled at atmospheric pressure, the gas liquefies to He I at 4.2 K. On further cooling a second-order transition to He II, another liquid form, occurs at 2.2 K. This form persists down to as close to 0 K as can be attained. To form solid helium, a pressure of 25 atm is required at 2.2 K. The mass of the ^{4}He atom is low and the intermolecular interactions are very weak. This means that the motion of the helium atoms is unusually large, too large in fact for a solid-like structure to persist except when high pressure forces closer distances and hence restricts the motion.

The density of He I at the boiling point at 1 atm is $125\,\mathrm{kg\,m}^{-3}$ and the viscosity is $\sim 3 \times 10^{-6}\,\mathrm{Pa\,s}$. As we would anticipate, cooling increases the viscosity until He II is formed. Cooling this form reduces the viscosity so that close to 0 K a liquid with zero viscosity is produced. The vibrational motion of the helium atoms is about the same or a little larger than the mean interatomic spacing and the flow properties cannot be considered in classical terms. Only a quantum mechanical description is satisfactory. We can consider this condition to give the limit of $De \rightarrow 0$ because we have difficulty in defining a relaxation when we have the positional uncertainty for the structural components.

3.5 MACROMOLECULAR FLUIDS

So far we have considered systems in which the flow units have been made up of only a few atoms. This is satisfactory when we consider fluids such as common solvents or simple lubricating oils. Many of the materials that we use every day are made up of very large numbers of atoms and we must recognise how this extends some of the problems that we have considered so far. These large aggregates of atoms and molecules may be in the form of particles or large soluble species. We need only think of items such as foods, agrochemicals, pharmaceuticals, paints and inks to have an abundance of examples. Although these systems all have some special problems due to their heterogeneity, they also have a strong element of the flow response governed by intermolecular forces. We will now consider colloidal dispersions, self-assembly surfactant systems and polymer systems.

3.5.1 Colloidal Dispersions

In this group of disperse systems we will focus on particles, which could be solid, liquid or gaseous, dispersed in a liquid medium. The particle size may be a few nanometres up to a few micrometres. Above this size the chemical nature of the particles rapidly becomes unimportant and the hydrodynamic interactions, particle shape and geometry dominate the flow. This is also our starting point for particles within the colloidal domain although we will see that interparticle forces are of great importance.

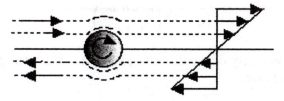

Figure 3.10   *The dilation of the flow field around a spherical particle. The shear field has a vorticity equal to $\dot{\gamma}/2$ and the particle rotates with this constant angular velocity*

3.5.2   Dilute Dispersions of Spheres

This is our starting point and the infinite dilution case was analysed by Einstein in the early years of this century.[12] This analysis was based on the dilation of the flow field because the liquid has to move around the flowing particle. The particles were assumed to be hard spheres so that they were rigid, uncharged and without attractive forces; small compared to any measuring apparatus so that the dilational perturbation of the flow would be unbounded and would be able to decay to zero (the hydrodynamic disturbances decay slowly with distance, *i.e.* r^{-1}); and at such dilution that the disturbance around one particle would not interact with the disturbance around another. The flow field is sketched in Figure 3.10. The coordinates are centred on the particle so that the symmetry is clear. The result of the analysis for slow flows (*i.e.* at low Reynolds number) was:

$$\eta = \eta_{\mathrm{o}}\left(1 + \frac{5\varphi}{2} + \mathrm{O}(\varphi^2)\right) \qquad (3.42)$$

where η is the dispersion viscosity, η_{o} is the viscosity of the dispersion medium and φ is the volume fraction and all terms in the volume fraction with an exponent greater than 1 are neglected. This limits the concentration to $\varphi < 0.01$. The coefficient 5/2 is known as the intrinsic viscosity, $[\eta]$, and will vary with the nature of the particle. For example if the assumption of rigidity is dropped, Taylor[13] showed that a fluid drop with a thin interface is deformed by the shear forces to give a prolate ellipsoid whose axial ratio is a function of the interfacial tension.[14] The orientation of the ellipsoid is along the principal axes of shear – *i.e.* at 45° to the direction of flow – and as the stress is transferred across the interface, circulation occurs inside the drop which results in *a decrease* in the rate of energy dissipation. The analysis gives the intrinsic viscosity for an uncharged emulsion drop as:

$$[\eta] = 2.5\left(\frac{\eta_i + 0.4\eta_o}{\eta_i + \eta_o}\right) \tag{3.43}$$

Here η_i is the viscosity of the disperse phase. It is interesting to look at various limiting cases:

$\eta_i = \infty$; $[\eta] = 2.5$; the Einstein value
$\eta_i = \eta_o$; $[\eta] = 1.75$
$\eta_i = 0$; $[\eta] = 1$; the situation for gas bubbles

This latter case is the same result as Einstein calculated for the situation where slip occurred at the rigid particle–liquid interface. Cox[15] has extended the analysis of drop shape and orientation to a wider range of conditions, but for typical colloidal systems the deformation remains small at shear rates normally accessible in the rheometer. The data shown in Figure 3.11 was calculated from Cox's analysis. His results have been confirmed by Torza *et al.*[16] with optical measurements. The ratio of the viscous to interfacial tension forces, R_f, was given as:

$$R_f = \frac{20\left(\frac{19}{16}\frac{\eta_i}{\eta_o} + 1\right)}{\left(1 - \frac{\eta_i}{\eta_o}\right)\left[\left(19\frac{\eta_i}{\eta_o}\right)^2 + \left(\frac{20\gamma_{io}}{\dot{\gamma}a\eta_o}\right)^2\right]^{\frac{1}{2}}} \tag{3.44}$$

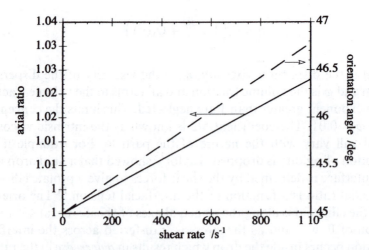

Figure 3.11 *The axial ratio and orientation angle of a liquid drop 10 μm in diameter and with $\eta_i = 0.8\eta_o$ and $\gamma_{io} = 40\,mN\,m^{-1}$ (data calculated from the analysis of Cox[15])*

where the drop radius is a and the interfacial tension is γ_{io}. Now the axial ratio of the prolate ellipsoid is

$$r_a = \frac{R_f + 1}{1 - R_f} \tag{3.45}$$

and its orientation angle in the shear field relative to the flow direction is

$$\alpha = \frac{\pi}{4} + \frac{1}{2}\tan^{-1}\left(\frac{19\,\dot{\gamma}a\eta_i}{20\,\gamma_{io}}\right) \tag{3.46}$$

As the concentration is increased above $\varphi \sim 0.01$, hydrodynamic interactions between particles become important. In a flowing suspension, particles move at the velocity of the streamline corresponding to the particle centre. Hence particles will come close to particles on nearby streamlines and the disturbance of the fluid around one particle interacts with that around passing particles. The details of the interactions were analysed by Batchelor[17] and we may write the viscosity in shear flow as

$$\frac{\eta(0)}{\eta_o} = 1 + 2.5\varphi + 6.2\varphi^2 + O(\varphi^3) \tag{3.47a}$$

or in extensional flow:

$$\frac{\eta(0)}{\eta_o} = 1 + 2.5\varphi + 7.6\varphi^2 + O(\varphi^3) \tag{3.47b}$$

There are three important points about Equation (3.47). Firstly the viscosity is the low shear limiting value, $\eta(0)$, indicating that we may expect some thinning as the deformation rate is increased. The reason is that a 'uniform' distribution was used (ensured by significant Brownian motion, *i.e.* $Pe < 1$) and this microstructure will change at high rates of deformation. Secondly there is a difference between the result for shear and that for extension. Thirdly the equation is only accurate up to $\varphi < 0.1$ as terms of order φ^3 become increasingly important. If we write the equation in the form often used for polymer solutions we have for Equation (3.47a):

$$\frac{\eta(0)}{\eta_o} = 1 + [\eta]\varphi + k_H[\eta]^2\varphi^2 + O(\varphi^3) \tag{3.48}$$

where the intrinsic viscosity for a hard sphere is $[\eta] = 2.5$ and the Huggins coefficient, k_H, is 1. The Huggins coefficient can now be thought of as an interaction parameter characterising the colloidal interactions between the particles as opposed to the purely hydrodynamic ones, *i.e.* either forces of repulsion or attraction will mean that $k_H > 1$. As the concentration is increased, interactions occurring between three or four particles become increasingly important. Unfortunately we have no rigorous analysis of this situation and so if the power series is extended to give terms of φ^3 and φ^4 the coefficients are unknown. Of course we could as a first 'guess' assume that we should see the same type of interaction between several particles as between two so:

$$\frac{\eta(0)}{\eta_o} = 1 + [\eta]\varphi + k_H([\eta]\varphi)^2 + k_H^2([\eta]\varphi)^3 + k_H^3([\eta]\varphi)^4 + \dots \qquad (3.49)$$

However, we should seek a more reliable solution that will describe the full range of volume fractions at which flow can occur and give some guidance as to the shear thinning behaviour.

3.5.3 Concentrated Dispersions of Spheres

There are numerous equations in the literature describing the concentration dependence of the viscosity of dispersions. Some are from curve fitting whilst others are based on a model of the flow. A common theme is to start with a dilute dispersion, for which we may define the viscosity from the hydrodynamic analysis, and then to consider what occurs when more particles are added to replace some of the continuous phase. The best analysis of this situation is due to Dougherty and Krieger[18] and the analysis presented here, due to Ball and Richmond,[19] is particularly transparent and emphasises the problem of excluded volume. The starting point is the differentiation of Equation (3.42) to give the initial rate of change of viscosity with concentration:

$$d\eta = 2.5\eta_o d\varphi \qquad (3.50)$$

Now consider a suspension at a volume fraction φ with a viscosity of $\eta(\varphi)$. When a small increase in concentration of this suspension is caused by replacing some of the medium by more particles, the expected increase in viscosity from the value η is

$$d\eta = 2.5\eta(\varphi)d\varphi \qquad (3.51)$$

Hence to calculate an increase in viscosity, Equation (3.51) must be integrated up to the value of interest so

$$\int \frac{d\eta}{\eta(\varphi)} = 2.5 \int d\varphi \qquad (3.52)$$

This 'effective medium' or 'mean field' assumption is easy to understand if there is a very large size difference between the newly added particles and any there previously, for example if we think of adding particles to a molecular liquid, we merely treat the liquid as a structureless continuum. However when the dimensions become comparable, the finite volume of the particles present prior to each addition must be considered, *i.e.* new particles can *only* replace medium and *not* particles. The consequence of this 'crowding' is that the concentration change is greater than expressed in Equation (3.52) and it must be corrected to the volume available:

$$\int \frac{d\eta}{\eta(\varphi)} = 2.5 \int \frac{d\varphi}{\left(1 - \dfrac{\varphi}{\varphi_m}\right)} \qquad (3.53)$$

Where φ_m is the maximum concentration at which flow is possible – above this solid-like behaviour will occur. φ/φ_m is the volume effectively occupied by particles in unit volume of the suspension and therefore is not just the geometric volume but is the *excluded volume*. This is an important point that will have increasing relevance later. Now integration of Equation (3.53) with the boundary condition that $\eta(\varphi) \to \eta_o$ as $\varphi \to 0$ gives

$$\frac{\eta(\varphi)}{\eta_o} = \left(1 - \frac{\varphi}{\varphi_m}\right)^{-2.5\varphi_m}$$

or more generally:

$$\frac{\eta(\varphi)}{\eta_o} = \left(1 - \frac{\varphi}{\varphi_m}\right)^{-[\eta]\varphi_m} \qquad (3.54)$$

There are two important issues concerning the factor that gives the excluded volume $1/\varphi_m$. These are (i) what is the effect of shear rate? (ii) what is the effect of polydispersity?

Few dispersions in everyday use are monodisperse and this will mean modification to Equation (3.54). A general expectation resulting from polydispersity is that denser packing may be achieved.[22] The simplest case occurs for bi- or multimodal systems with very large size differences between each mode, several orders of magnitude for example. To calculate the zero shear viscosity of systems containing 1, 1:3, and

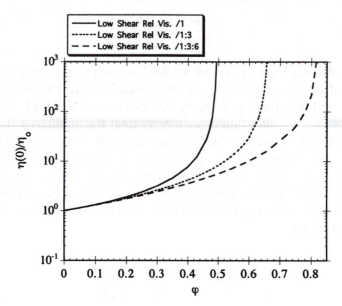

Figure 3.12   *The low shear limiting viscosity for unimodal, bimodal and trimodal size distributions calculated from Equation (3.55)*

$1:3:6$ size ratios of large to small particles as a function of volume fraction we consider the contribution of the volume fraction of each size fraction in turn. Then each addition of particles is to an effective medium and Equation (3.55) becomes the product relationship:

$$\frac{\eta(0)}{\eta_o} = \Pi_n \left(1 - \frac{\varphi_n}{\varphi_m} \right)^{-1.24} \tag{3.55}$$

where there are n narrow modes in the distribution. The result is shown in Figure 3.12. The effect of the ratio of large to small particles by volume in a bimodal mix is plotted in Figure 3.13. However we must remember that these results are for the case where the size ratios are very large. When this is not the case, the excluded volume of the larger particles will tend to increase as the packing efficiency is reduced. There is, however, an effect of the shear rate. Under quiescent conditions, particles in a dispersion move around with Brownian motion. Computer simulation of hard spheres[20,21] indicates that as the concentration is increased a liquid to solid or freezing transition occurs at $\varphi = 0.495$ (and that on dilution a melting transition commences at $\varphi = 0.54$), *i.e.* $[\eta]\varphi_m = 1.24$. The effect of high shear rates is to organise the particles into a layered or string-like arrangement because shear field dominates the Brownian motion. This structure was first shown by Hoffman[23] and has been confirmed by

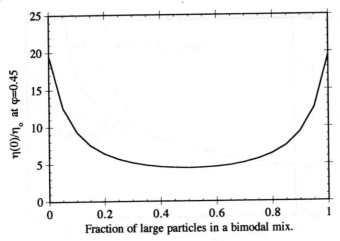

Figure 3.13 *The effect on the viscosity of the volume ratio in a bimodal mixture of spherical particles*

optical rheometry, neutron scattering and computer simulation. The upper limit for φ can therefore be set at the packing of monodisperse spheres in layers with hexagonal symmetry, hence for this density of packing $\varphi = 0.605$, *i.e.* $[\eta]\varphi_m = 1.51$. Equation (3.54) may now be rewritten for monodisperse hard spheres as

$$\frac{\eta(0)}{\eta_o} = \left(1 - \frac{\varphi}{0.495}\right)^{-1.24} ; \quad \frac{\eta(\infty)}{\eta_o} = \left(1 - \frac{\varphi}{0.605}\right)^{-1.51} \tag{3.56}$$

The results of Equation (3.56) are plotted in Figure 3.14. It can be seen that shear thinning will become apparent experimentally at $\varphi > 0.3$ and that at values of $\varphi > 0.5$ no zero shear viscosity will be accessible. This means that solid-like behaviour should be observed with shear melting of the structure once the yield stress has been exceeded with a stress controlled instrument, or a critical strain if the instrumentation is a controlled strain rheometer. The most recent data[24,25] on model systems of nearly hard spheres gives values of maximum packing close to those used in Equation (3.56).

The shear rate range over which the shear thinning of hard sphere suspensions occurs can be determined from the equations due to Krieger[26] or Cross[27] in conjunction with the reduced stress or Péclet number respectively:

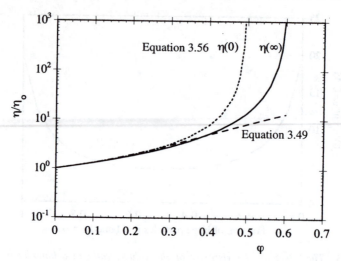

Figure 3.14   *The relative viscosity as a function of volume fraction from Equations (3.49) and (3.56)*

Krieger:

$$\eta(\sigma_r) = \eta(\infty) + \frac{(\eta(0) - \eta(\infty))}{1 + \left(\dfrac{\sigma_r}{\sigma_c}\right)^n} \tag{3.57a}$$

Cross:

$$\eta(Pe) = \eta(\infty) + \frac{(\eta(0) - \eta(\infty))}{1 + \left(\dfrac{Pe}{Pe_c}\right)^m} \tag{3.57b}$$

where the reduced stress σ_r is given by

$$\sigma_r = \frac{\sigma a^3}{k_B T} \tag{3.58}$$

Note that $\sigma_r = \dot{\gamma}\eta(\sigma_r)a^3/k_B T$ and recall that in a concentrated dispersion the Péclet number is $Pe = 6\pi\dot{\gamma}\eta(\sigma_r)a^3/k_B T$. The use of the suspension viscosity implies that the particle diffusion can be estimated from an effective medium approach. Both Krieger and Cross gave the power law indices (n and m) as 1 for monodisperse spherical particles. In this formulation, the subscript c indicates the characteristic value of the reduced stress or Péclet number at the mid-point of the viscosity curve. The expected value of Pe_c is 1, as this is the point at which diffusional and convective timescales are equal. This will give a value of $\sigma_c \sim 5 \times 10^{-2}$. Figure 3.15 shows a plot of Equation (3.57a) with this value and $n = 1$

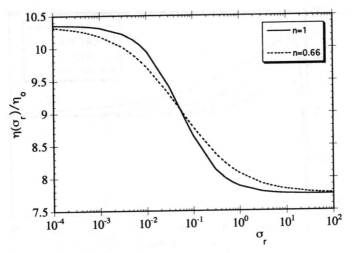

Figure 3.15 *Calculated shear thinning response for a hard sphere dispersion at $\varphi = 0.45$*

and 0.66. Cross indicated that for increasing polydispersity m would approach 2/3.

3.5.4 Charge Stabilised Dispersions

The use of surface charge to provide colloid stability to particles dispersed in dilute electrolytes in aqueous solution, or even in media of intermediate polarity, is an effective means of stabilising particles against van der Waals' forces of attraction. Figure 3.16 shows typical potential

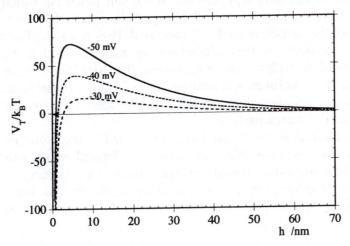

Figure 3.16 *The pair potential for rutile in ethylene glycol at infinite dilution as a function of diffuse layer potential. Background concentration 1×10^{-4} M 1 : 1 electrolyte*

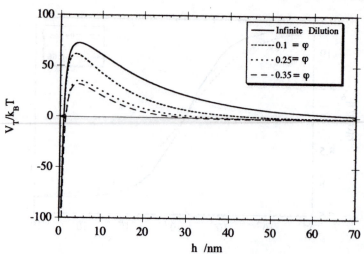

Figure 3.17   *The pair potential as a function of volume fraction due to diffuse layer ions. Diffuse layer potential = −50 mV and background electrolyte concentration 1 × 10⁻⁴ M*

energy curves calculated for titanium dioxide dispersions in ethylene glycol at different values of the diffuse layer potential. The pair potential can be calculated from the usual DLVO theory as outlined in Chapter 2. The increase in particle concentration means that the total ion concentration increases, and this also contributes to the screening of the electrostatics, in the manner illustrated in Figure 3.17. The range of the interaction is an important feature when considering the viscosity of suspensions. The long repulsive tail of the pair potential, which is of a similar order as the size of a colloidal particle, means that the excluded volume of the particle is markedly increased. However from Figures 3.16 and 3.17 it is clear that the excluded volume will be a function of both the background electrolyte concentration and the particle concentration. In addition the shear forces will push the particles closer together and hence the excluded volume will be smaller at higher shear rates, in addition to any structural reorganisation.

Three electroviscous effects have been noted in the literature.[27] The primary electroviscous effect refers to the enhanced energy dissipation due to the distortion of the diffuse layer from spherical symmetry during flow. The analysis for low diffuse layer potentials has been clearly reviewed by van de Ven[28] and the result for the intrinsic viscosity with $\kappa a \to \infty$ is:

$$[\eta] = 2.5 + \frac{6(\varepsilon\zeta)^2}{K\eta_o a^2} \tag{3.59}$$

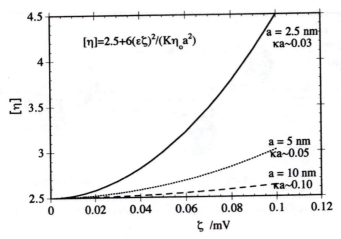

Figure 3.18 *The primary electroviscous effect estimated from Equation (3.59) for spherical particles dispersed in 10^{-2} M KCl*

where K is the specific conductivity.

Figure 3.18 shows how this varies as the ζ-potential is varied at an electrolyte concentration of 10^{-2} M potassium chloride. Once the potential exceeds 50 mV the calculation error becomes large and Equation (3.59) should really be restricted to $\kappa a > 100$. However, Figure 3.18 shows how small the effect normally is. A numerical solution valid for any potential and all κa values was published by Watterson and White.[29]

The secondary electroviscous effect is the enhancement of the viscosity due to particle–particle interactions, and this of course will control the excluded volume of the particles. The most complete analysis is that due to Russel[30] and we may take this analysis for pair interactions as the starting point. Russel's result gives the viscosity as

$$\frac{\eta(\dot{\gamma})}{\eta_0} = 1 + \frac{5}{2}\varphi + \left(\frac{5}{2} + \frac{3}{40}\left(\frac{r_0}{a}\right)^5\right)\varphi^2 + \ldots \tag{3.60}$$

Now r_0 is the minimum centre-to-centre separation of a colliding pair of particles. The value of r_0 may be estimated from the balance of the electrostatic repulsion and the forces bringing the particle together (*i.e.* Brownian motion and shear field). The limiting value at low shear will be given by Brownian motion probing the repulsion and, at higher shear, the hydrodynamic forces dominate and these may be estimated from the radial component of the Stokes' drag force. Figure 3.19 shows the particle trajectories during collision, drawn with curvilinear shapes as a simplifying approximation.

In the low shear limit, the Brownian interaction is the dominant

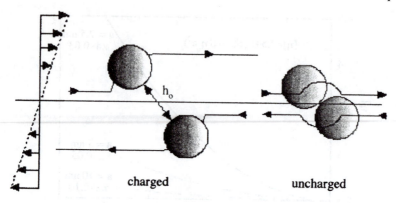

Figure 3.19 *The collision trajectories of particles in a shear field,* $r_0 = h_o + 2a$

component. Only the electrostatic repulsion needs to be considered if the particles are small and the electrolyte concentration is low, *i.e.* the van der Waals' attraction is much less than the electrostatic repulsion. The problem is to determine the average minimum separation of the particles produced by Brownian motion and so we can say

$$\frac{V_e(r)}{k_B T} \approx 1 \text{ at } r_0(0) \tag{3.61}$$

where

$$\frac{V_e(r)}{k_B T} = \left(\frac{4\pi\varepsilon\kappa(a\zeta)^2 \exp(2\kappa a)}{k_B T} \right) \frac{\exp(-\kappa r)}{\kappa r} = \alpha \frac{\exp(-\kappa r)}{\kappa r}$$

so that the asymptotic approximation to the effective hard sphere diameter is

$$r_0(0) \sim \frac{1}{\kappa} \ln\{\alpha / \ln[\alpha / \ln(\alpha / \dots)]\} \tag{3.62}$$

Substitution of Equation (3.62) into Equation (3.60) gives the relative zero shear viscosity. When the shear rate makes a significant contribution to the interparticle interactions, the mean minimum separation can be estimated from balancing the radial hydrodynamic force, F_{hr}, with the electrostatic repulsive force, F_e. The maximum radial forces occur along the principle axes of shear, *i.e.* at an orientation of the line joining the particle centres to the streamlines of $\theta = 45°$. This is the orientation shown in Figure 3.19. The hydrodynamic force is calculated from the Stokes drag, $6\pi\eta_o au$, where u is the particle velocity, which is simply

$u = \dot{\gamma} r_o \sin\theta$. So decomposing this into the tangential and radial components gives:

$$F_{hr} = 6\pi\eta_o\dot{\gamma}r_o \sin^2\theta = 3\pi\eta_o\dot{\gamma}r_o \qquad (3.63)$$

and the electrostatic repulsive force is

$$F_e = 4\pi\varepsilon(a\zeta)^2 \exp(2\kappa a)\frac{1+\kappa r_o}{r_o^2}\exp(-\kappa r_o) \qquad (3.64)$$

From balancing these two forces we may write

$$\frac{1+\kappa r_o}{r_o^3}\exp(-\kappa r_o) = \frac{3\eta_o\dot{\gamma}}{4\varepsilon a\zeta^2\exp(2\kappa a)} \qquad (3.65)$$

The results from Equation (3.65) are plotted in Figure 3.20 for 100 nm particles with a ζ-potential of 50 mV dispersed in 10^{-5} and 10^{-3} M sodium chloride solutions. From the graph it can be seen that at low shear ($< 30\,\text{s}^{-1}$) the particles have an effective hard sphere diameter of ~ 650 nm for the lowest electrolyte concentration and this only approaches the particle diameter at shear rates $> 10^4\,\text{s}^{-1}$. Thus in the low shear limit the particles only approach each other to a distance of $\sim 4.5\kappa^{-1}$. At the higher electrolyte concentration, the effective hard sphere diameter is much closer to the particle diameter at ~ 250 nm and the greater steepness of the repulsive force curve is shown in the range of shear rates required to vary this dimension.

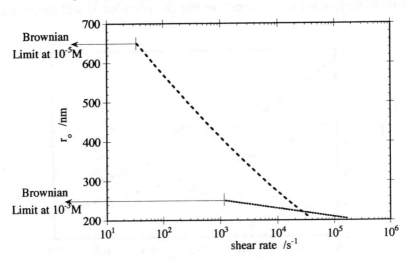

Figure 3.20 *The effective hard sphere diameter, r_o, calculated from Equation (3.65) for 100 nm radius particles with $\zeta = 50\,mV$*

The effective hard sphere diameter may be used to estimate the excluded volume of the particles, and hence the low shear limiting viscosity by modifying Equation (3.56). The liquid/solid transition of these charged particles will occur at

$$\varphi_{me} = 0.495 \left(\frac{2a}{r_o} \right)^3 \tag{3.66}$$

However some problems must be addressed prior to the use of Equation (3.66) from the data generated above. The first problem is that the ions in the diffuse layers must be included in the estimation of κ. This becomes essential as soon as the particle–particle interactions become significant and so Equations (3.62) to (3.65) should contain a volume fraction dependence of κ of the form given by Russel,[30] for example for a symmetrical electrolyte:

$$\kappa = \sqrt{\frac{e^2}{\varepsilon k_B T} \frac{2z^2 n_o - \frac{3qz\varphi}{ea}}{1 - \varphi}} \tag{3.67}$$

Here z is the ion valency, n_o is the number density of added electrolyte and q is the surface charge density of the particles. Figure 3.21 clearly illustrates the sensitivity at the lower added electrolyte concentrations where the diffuse layer Debye length is equivalent to that which would be estimated for an added electrolyte concentration two orders of magnitude higher, at a volume fraction of 0.5. Clearly higher concentrations of added electrolyte are not as sensitive but the variation is still significant.

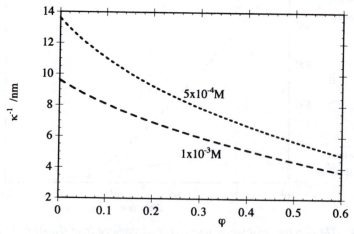

Figure 3.21 *The variation of the Debye length with volume fraction of particles*

The next problem is that the variation of r_o with shear rate is only valid as written in Equation (3.65) at low volume fractions because the solvent viscosity is used to calculate the value of r_o over the shear rate range. If an effective medium treatment is used to make a simple estimate of the effect of many-body hydrodynamic interactions we have:

$$\frac{1 + \kappa r_o(\dot{\gamma})}{r_o^3(\dot{\gamma})} \exp[-\kappa r_o(\dot{\gamma})] = \frac{3\eta(\dot{\gamma})\dot{\gamma}}{4\varepsilon a \zeta^2 \exp(2\kappa a)} = \frac{3\sigma(\dot{\gamma})}{4\varepsilon a \zeta^2 \exp(2\kappa a)} \qquad (3.68)$$

The calculation is not straightforward because we need the effective hard sphere diameter at each shear rate in order to estimate the viscosity at that shear rate and we also need to consider the concomitant structural reorganisation. In summary, we may easily define the zero shear viscosity. This is plotted in Figure 3.22 for a polystyrene latex dispersion in 1×10^{-3} and 5×10^{-4} M sodium chloride solutions. The mean particle size was $a = 85\,nm$ and the ζ-potential was measured at $78\,mV$. The calculated curves fit reasonably well to the experimental points. The high shear limiting viscosity, calculated from Equation (3.56), is shown for comparison. It is clear from Figure 3.22 that the liquid/solid transition has moved to much lower volume fractions and is markedly dependent on the electrolyte concentration. The rheological behaviour that we observe experimentally at 5×10^{-4} M is (1) for $0 < \varphi < 0.1$, the dispersion is close to Newtonian; (2) at concentrations $0.1 < \varphi < 0.21$ the dispersion is pseudoplastic and both high and low shear viscosities are

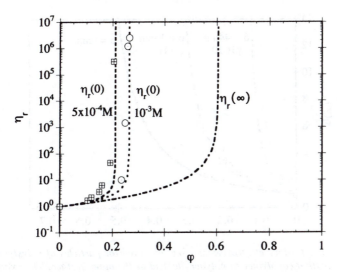

Figure 3.22 *The high and low shear viscosity of a polystyrene latex with $a = 85\,nm$, $\zeta = 78\,mV$ in 5×10^{-4} and 1×10^{-3} M NaCl*

experimentally accessible; and (3) for $0.21 < \varphi$, solid-like behaviour is observed with flow resulting from the shear melting of the system.[31]

The tertiary electroviscous effect is the change in viscosity of a polymer solution or a dispersion due to the change in conformation of polyelectrolytes as the pH or ionic strength is changed. The charged groups along the polyelectrolyte chain repel each other. A typical example of such a system is poly(acrylic acid). The weak acid groups interact strongly with each other because they are only separated by two carbon atoms. This means that the pK_a of the polyelectrolyte is no longer a single value and full dissociation of the groups is not achieved until pH values above 8 are reached. Electrolytes screen the interactions and hence the conformation of the polymer in solution is a function of the combination of the pH and salinity.

Polyelectrolytes provide excellent stabilisation of colloidal dispersions when attached to particle surfaces as there is both a steric and electrostatic contribution, *i.e.* the particles are 'electrosterically' stabilised. In addition the origin of the electrostatic interactions is displaced away from the particle surface and the origin of the van der Waals' attraction, reinforcing the stability. Kaolinite stabilised by poly(acrylic acid) is a combination that would be typical of a paper-coating clay system. Acrylic acid or methacrylic acid is often copolymerised into the latex particles used in cement sytems giving particles which swell considerably in water. Figure 3.23 illustrates a viscosity curve for a copoly(styrene–

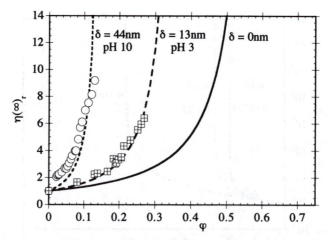

Figure 3.23   *The tertiary electroviscous effect observed for particles of polystyrene latex with a copolymer of polyacrylic acid at the outer surface. The experimental points were obtained at pH 3 and 10. The dry particle radius was 75 nm and $\kappa a \sim 25$*

acrylic acid) latex system where the particle dimensions are very sensitive to the value of pH.

3.6 REFERENCES

1. R.B. Bird, W.E. Stewart and E.N. Lightfoot, in *Transport Phenomena*, Wiley, New York, 1960, p. 211.
2. G.I. Taylor, *Phil. Trans.* 1923, **A223**, 289.
3. See for example, P.W. Atkins, *Physical Chemistry*, 4th edn, Oxford University Press, Oxford, 1990.
4. D. Tabor, in *Gases, Liquids and Solids*, 3rd edn, Cambridge University Press, Cambridge, 1991, p. 66.
5. G. Harrison, in *The Dynamic Properties of Supercooled Liquids*, Academic Press, London, 1976, p. 13.
6. F.A. Lindemann, *Z. Phys.* 1910, **11**, 609.
7. R.C. Weast (Ed.), *The Handbook of Chemistry and Physics*, 53rd edn, CRC Press, Ohio, 1972.
8. A. Kovacs, *J. Polymer Sci.* 1958, **30**, 131.
9. A.K. Doolittle, *J. Appl. Phys.* 1951, **22**, 1471.
10. J. Hildebrand, *J. Chem.Soc., Faraday Disc.* 1978, **66**, 151.
11. H. Eyring, *J. Chem. Phys.* 1936, **4**, 283.
12. A. Einstein, *Ann. Physik.* 1906, **19**, 289; 1911, **34**, 591.
13. G.I. Taylor, *Proc. R. Soc. London* 1932, **A138**, 41.
14. G.I. Taylor, *Proc. R. Soc. London* 1934, **A146**, 501.
15. R.G. Cox, *J. Fluid Mech.* 1969, **37**, 601.
16. S. Torza, R.G. Cox and S.G. Mason, *J. Colloid Interface Sci.* 1972, **38**, 395.
17. G.K. Batchelor, *J. Fluid Mech.* 1977, **83**, 97.
18. I.M. Krieger and T.J. Dougherty, *Trans. Soc. Rheol.* 1959, **3**, 137.
19. R. Ball and P. Richmond, *Phys. Chem. Liquids* 1980, **9**, 99.
20. B.J. Alder and T.E. Wainwright, *Phys. Rev.* 1962, **127**, 359.
21. W.G. Hoover and F.H. Ree, *J. Chem. Phys.* 1968, **49**, 3609.
22. R.M. German, *Particle Packing Characteristics*, Metal Powder Industries Federation, New York, 1989, p. 120.
23. R.L. Hoffman, *J. Coll. Interface Sci.* 1974, **46**, 491.
24. R.A. Lionberger and W.B. Russel, *J. Rheol.* 1997, **41**, 399.
25. W.J. Frith, P. d'Haene, R. Buscall and J. Mewis, *J. Rheol.* 1996, **40**, 531.
26. I.M. Krieger, *Adv. Coll. Interface Sci.* 1972, **3**, 111.
27. M.M. Cross, *J. Coll. Sci.* 1965, **20**, 417; 1970, **33**, 30.

CHAPTER 4

Linear Viscoelasticity I. Phenomenological Approach

4.1 VISCOELASTICITY

Many materials can be readily classified as solids or fluids, displaying elastic and viscous behaviour, respectively. As with all such simple classification schemes it is far from complete. For example, fluids encompass both gases and liquids which show differing short-range molecular order. The thermodynamic properties of these systems are very different. The ideal method of assessing whether a material is a solid or a fluid is by experimental observation. However, whether a material appears as a solid or a liquid depends upon the length of time over which you are prepared to make observations. As an example consider lead, which is of course a metallic solid and used in the past as a roofing material for churches. However, lead possesses some fluid-like character and will creep and flow under its own weight. This happens slowly, over many years, but if a roof is pitched too steeply it could eventually slip completely off! Thus over a long enough timescale lead will appear to flow. Apparently fluid materials such as water systems, when subjected to short timescale ultrasonic pulses, display elastic behaviour, which is the principle underlying medical ultrasound scans. A classification of materials should include a consideration of the timescale of the measurement relative to the characteristic time of the material. The ratio of these times is given by the Deborah number (Chapter 1). When the Deborah number is of order 1 the material will display both viscous and elastic behaviour and is described as viscoelastic. In common with thermodynamic treatments, the viscoelastic response of a material can be presented in two complementary ways. First is a macroscopic viewpoint in which the material can be considered as a continuum. Its properties are considered to be dynamic but not explicitly ascribed to a microscopic mechanism. The properties depend upon the balance of energy storage and loss processes. These are used to determine a viscoelastic response, the value of which depends on the time dependence of the sample and the measurement timescale. Secondly, a microscopic picture of a sample is constructed, for example using a cell model or a thermodynamic ensemble. These microscopic models can be modified to allow for the application of a body force and more or less complex statistical mechanical models are used to generate the experimental response. This can be obtained from an analytical model or a computer simulation. Microscopic and macroscopic approaches are complementary. In this chapter a macroscopic view will be used to describe the phenomena arising from the application of a stress or strain to a body. Such an approach is described as a *phenomenological approach*.

4.2 LENGTH AND TIMESCALES

The phenomenological approach does not preclude a consideration of the molecular origins of the characteristic timescales within the material. It is these timescales that determine whether the observation you make is one which 'sees' the material as elastic, viscous or viscoelastic. There are great differences between timescales and length scales for atomic, molecular and macromolecular materials. When an instantaneous deformation is applied to a body the particles forming the body are displaced from their normal positions. They diffuse from these positions with time and gradually dissipate the stress. The diffusion coefficient relates the distance diffused to the timescale characteristic of this motion. The form of the diffusion coefficient depends on the extent of ordering within the material.

For an ideal gas the diffusion coefficient is related to the mean free path of the gas molecule, λ, which represents the mean distance between collisions for that molecule.

$$D = \frac{1}{3}\lambda\bar{c} \tag{4.1}$$

For a molecule at RTP this is of the order of a few hundred molecular diameters. In our ideal gas there is a distribution of velocities of the molecules about a mean value $\bar{c}$. The mean free path defines a length scale in gases. As the density of the gas is increased and the mean free path approaches the molecular dimensions, a short-range molecular order develops and the material condenses to a liquid. The diffusional length scale is now much shorter range as a molecule encounters its nearest neighbour on length scales approaching its molecular diameter. By treating all the molecules surrounding any test molecule of radius a, as a continuum with a viscosity η_0 the Stokes–Einstein equation can be derived (Section 1.3.2):

$$D = \frac{k_B T}{6\pi\eta_0 a} \tag{4.2}$$

In contrast to a gas, the short-range order in the fluid demands that significant structural relaxation must have occurred when the molecule has diffused a distance equivalent to the distance to the surrounding shell of nearest neighbours. This is of the order of a molecular radius. Since the diffusive process is described by the square of the mean distance moved by a molecule in a time τ then

$$\tau = \frac{6\pi\eta_0 a^3}{k_B T} \tag{4.3}$$

This expression represents the structural relaxation time of a liquid so that if a strain is applied to the material it will relax the stress with a time characterised by τ. This prompts the question 'What is the form of the stress when a strain is applied?' To answer this question we must consider linear viscoelasticity in detail.

4.3   MECHANICAL SPECTROSCOPY

The use of infra-red or ultraviolet spectroscopy to examine the molecular groups present in a chemical compound is familiar to any chemist. One of the main uses of this technique is to apply a range of electromagnetic frequencies to a sample and thus identify the frequency at which a process occurs. This can be characteristic of, say, the stretch of a carbonyl group or an electronic transition in a metal complex. The frequency, wavelength or wavenumber at which an absorption occurs is of most interest to an analytical chemist. In order to use this information quantitatively, for example to establish the concentration of a molecule present in a sample, the Beer–Lambert law is used:

$$I_{rel} = \exp(-\varepsilon_x cl) \tag{4.4}$$

Here the relative intensity I_{rel} of radiation being transmitted, travelling a distance l through a sample, has an exponential dependence on two properties, the concentration of the material c and ε_x, the extinction coefficient or absorption coefficient of the material. The value of ε_x

depends upon the nature of the transition (electronic, vibrational, *etc.*). Often this relationship is used at a particular wavelength, characteristic of a process in a material, and the magnitude of the absorbance indicates the amount of material present. An analogous mechanical measurement can be made by applying an oscillating strain at a range of frequencies. This will result in a stress in the sample with a magnitude characteristic of the material and its concentration. This will vary in magnitude as a function of frequency. An important difference between mechanical and electromagnetic probes is that the origin of the mechanical response is often very difficult to isolate as a single structural process. We must acquaint ourselves with the idea that viscoelastic responses often arise from a distribution of processes. It is also true that two different samples can have very similar rheological features but that they may be due to entirely different causes. It is these features that lead to the notion of a phenomenological treatment of the rheological behaviour. In this approach the origin of the observed process is not accounted for in terms of molecular or colloidal interactions but in terms of a combination of Hookean springs and Newtonian dashpots. These can be added together in series or in parallel, in more or less complex combinations. They are able to mimic the linear viscoelastic behaviour of any material.

4.4   LINEAR VISCOELASTICITY[1,2,3]

Modern instrumentation is capable of applying very complex stress or strain profiles to a sample. For example if you choose to apply an oscillating stress, the result is an oscillating strain. When you are in a linear viscoelastic region and you double the applied stress the resulting strain will also double. The ratio of the maximum stress to the maximum strain is constant. You will also observe this behaviour if you apply the stress to a Hookean spring or dashpot containing a Newtonian oil as shown in Figures 4.1 and 4.2.

A Hookean spring is one that obeys Hooke's law, *i.e.* the force is

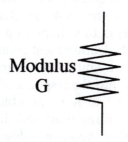

Figure 4.1   *A spring with modulus G*

Viscosity
η

Figure 4.2 *A dashpot containing an oil of viscosity η*

linearly related to the displacement. The response of the spring can be described by a shear modulus G. A Newtonian dashpot consists of a cup filled with a Newtonian oil of viscosity η, with a piston placed in the oil. Combinations of such devices are often used to dampen vibrations. It is of course possible for us to visualise situations in which the piston is pulled free from the oil or where the spring stretches too far and is permanently deformed. This would not be linearly viscoelastic! We are only interested in small strains or stresses where these two simple mechanical models display the property of linear viscoelasticity, where the piston remains in the oil or the spring recoils completely once the deformation is released. It is also true that any combination of these models will display linear viscoelasticity. It is this feature that gives these models their utility because seemingly complex experimental behaviour can be described by a combination of these simple models. This will enable us to relate the complex response of a system in terms of some simple parameters such as G and η. These in turn can be related via microstructural models to the main chemical features of the system.

4.4.1 Mechanical Analogues

The use of mechanical models such as the spring and dashpot as analogues of the behaviour of real materials enables us to describe very complex experimental behaviour using a simple combination of models. We can also use models to represent a wide range of deformations, so we need not restrict ourselves to the application of a shear stress. Models can be appropriately applied to extension or bulk deformations with the appropriate substitutions for the bulk or elongational moduli and viscosity. We will concentrate in this chapter on the shear properties of the material because this is the feature most commonly measured in the chemical laboratory. Ideally we would like our mechanical analogues to be able to describe more than just the response of our material to say an oscillating strain. We would like it to be able to predict the stress response to all the possible deformations and timescales we can apply, in other words to be a complete rheological description of the material. Such an expression is described as the *constitutive equation* of the body.

Whilst obtaining this is the ultimate goal for many rheologists, in practice it is not possible to develop such an expression. However, our mechanical analogues do allow us to develop linear constitutive equations which allow us to relate the phenomena of linear viscoelastic measurements. For a spring the relationship is straightforward. When any form of shear strain is placed on the sample the shear stress responds instantly and is proportional to the strain. The constant of proportionality is the shear modulus

$$\sigma = G\gamma \tag{4.5}$$

where σ is the stress due to the applied strain γ. This model has the response of a purely elastic solid. The constitutive equation is equally simple for the dashpot. When any form of shear strain rate is placed on the sample the shear stress responds instantly and is proportional to the strain rate. The constant of proportionality is the shear viscosity

$$\sigma = \eta\dot{\gamma} \tag{4.6}$$

Here the time derivative of the strain is represented by Newton's dot. This is the response of a purely viscous fluid. Now suppose we consider a combination of these models. The two simplest arrangements that we can visualise is the models in series or parallel. When they are placed in series we have a Maxwell model and in parallel we have a Kelvin (or sometimes a Kelvin–Voigt) model.

The constitutive equations for these models become more complex. When we apply a stress to the parallel elements of a Kelvin model both elements will respond. Thus a linear addition of the stresses describes the constitutive equation:

$$\sigma = G\gamma + \eta\dot{\gamma} \tag{4.7}$$

For a Maxwell model it is the strain rates that linearly add

$$\dot{\gamma} = \frac{\dot{\sigma}}{G} + \frac{\sigma}{\eta} \tag{4.8}$$

Both of these models show contributions from the viscosity and the elasticity, and so both these models show viscoelastic behaviour. You can visualise a more complex combination of models possessing more complex constitutive equations and thus able to describe more complex rheological profiles.

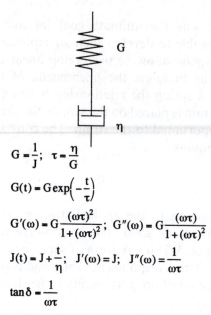

$$G = \frac{1}{J}; \quad \tau = \frac{\eta}{G}$$

$$G(t) = G \exp\left(-\frac{t}{\tau}\right)$$

$$G'(\omega) = G\frac{(\omega\tau)^2}{1+(\omega\tau)^2}; \quad G''(\omega) = G\frac{(\omega\tau)}{1+(\omega\tau)^2}$$

$$J(t) = J + \frac{t}{\eta}; \quad J'(\omega) = J; \quad J''(\omega) = \frac{1}{\omega\tau}$$

$$\tan\delta = \frac{1}{\omega\tau}$$

Figure 4.3 *A Maxwell model*

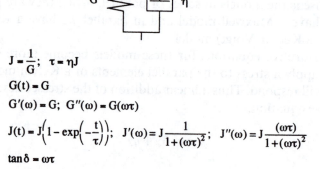

$$J = \frac{1}{G}; \quad \tau = \eta J$$

$$G(t) = G$$

$$G'(\omega) = G; \quad G''(\omega) = G(\omega\tau)$$

$$J(t) = J\left(1 - \exp\left(-\frac{t}{\tau}\right)\right); \quad J'(\omega) = J\frac{1}{1+(\omega\tau)^2}; \quad J''(\omega) = J\frac{(\omega\tau)}{1+(\omega\tau)^2}$$

$$\tan\delta = \omega\tau$$

Figure 4.4 *A Kelvin–Voigt model*

4.4.2 Relaxation Derived as an Analogue to First-Order Chemical Kinetics

The simplest experiment we can visualise performing is the very rapid application of a small strain which is then maintained at a constant level. This is simply a strain-jump experiment and can be treated in an analogous fashion to a temperature or a pressure-jump experiment. The stress will follow the strain and increase to a maximum value. For an

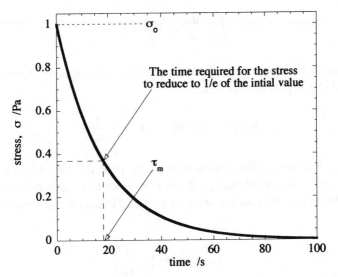

The time required for the stress to reduce to 1/e of the intial value

Figure 4.5 *A stress relaxation curve*

elastic material represented by a spring the stress will remain constant all the time the stress is applied. If the strain is applied to a Maxwell model you can visualise the spring responding instantly, giving rise to a stress. However, the stress in the spring will gradually reduce with time, as the piston slowly moves through the fluid in the dashpot. The reduction in stress can be considered to be analogous to a first-order rate process. The curve describing the decay of stress with time is shown in Figure 4.5.

It is a monotonically decreasing function with time, where the rate of decay is directly proportional to the stress. Using a first-order rate expression to describe the decay of stress we get the following expression:

$$\frac{\mathrm{d}\sigma}{\mathrm{d}t} = -k\sigma \tag{4.9}$$

where k represents the rate constant. We can rearrange the expression to gather all the terms in stress on the left-hand side:

$$\frac{\mathrm{d}\sigma}{\sigma} = -k\mathrm{d}t \tag{4.10}$$

Now in order to obtain an expression for stress in terms of time we need to integrate from the initial time $t = 0$, where the maximum stress $\sigma(0)$ is achieved, to a time t and stress $\sigma(t)$ later:

$$\int_{\sigma(0)}^{\sigma(t)} \frac{d\sigma}{\sigma} = -\int_0^t k\,dt \tag{4.11}$$

Evaluating this expression gives an explicit relationship between the stress and the experimental time t:

$$\ln \sigma(t) - \ln \sigma(0) = -kt = -\frac{t}{\tau_m} \tag{4.12}$$

The rate constant has the dimensions of reciprocal time and so k can be replaced by a time constant τ_m. If we now take the exponential of both sides of this expression and rearrange we can obtain an expression for the stress:

$$\sigma(t) = \sigma(0) \exp\left(-\frac{t}{\tau_m}\right) \tag{4.13}$$

This expression describes the decay of the stress with time when a rapid strain is applied to a Maxwell model. The characteristic decay time τ_m is given by the ratio of the viscosity to the shear modulus:

$$\tau_m = \frac{\eta}{G} \tag{4.14}$$

Now if we divide the shear stress in Equation (4.13) by the applied strain we obtain an expression in the form of a shear modulus. This term $G(t)$ is described as the relaxation function:

$$G(t) = \frac{\sigma(t)}{\gamma} = \frac{\sigma(0)}{\gamma} \exp\left(-\frac{t}{\tau_m}\right) \tag{4.15}$$

At very short experimental times compared with τ_m the exponential term tends to 1. Under these circumstances the relaxation function tends to the value of the modulus of the spring. The response is simply that of the spring so that the initial stress divided by the strain gives the modulus of the spring.

$$G(t) = G \exp\left(-\frac{t}{\tau_m}\right) \tag{4.16}$$

This expression describes the decay of the shear modulus with time. The very short time response is that of the spring:

$$G(t \to 0) = G \tag{4.17}$$

The spring is elastically storing energy. With time this energy is dissipated by flow within the dashpot. An experiment performed using the application of rapid stress in which the stress is monitored with time is called a stress relaxation experiment. For a single Maxwell model we require only two of the three model parameters to describe the decay of stress with time. These three parameters are the elastic modulus G, the viscosity η and the relaxation time τ_m. The exponential decay described in Equation (4.16) represents a linear response. As the strain is increased past a critical value this simple decay is lost.

4.4.3   Oscillation Response

In this section we deal with perhaps the most conceptually difficult of all the responses observed in linear viscoelastic materials. This is the response of a material to an oscillating stress or strain. This is an area that illustrates why rheological techniques can be considered as mechanical spectroscopy. When a sample is constrained in, say, a cone and plate assembly, an oscillating strain at a given frequency can be applied to the sample. After an initial start-up period, a stress develops in direct response to the applied strain due to transient sample and instrumental responses. If the strain has an oscillating value with time the stress must also be oscillating with time. We can represent these two wave-forms as in Figure 4.6.

All the information about the response of the sample at this frequency

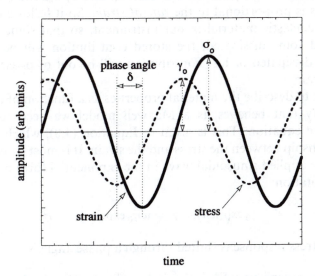

Figure 4.6   *An oscillating strain and the stress response for a viscoelastic material*

is contained within these wave-forms. However this information is not in a particularly tractable form. What we would really prefer is to have a few representative terms such as the relaxation time and elasticity or viscosity of the sample in order to characterise its material properties. In order for us to obtain this information some mathematical manipulation is required. Two key features can be utilised, which are constant in time for any given frequency. The first feature is that the maximum stress, σ_0, divided by the maximum strain, γ_0, is constant for a given frequency ω. The ratio of these terms is called the complex modulus G^*:

$$|G^*(\omega)| = \frac{\sigma_0}{\gamma_0} \tag{4.18}$$

The term ω is the radial frequency, which is equal to $2\pi f$ where f is the applied frequency measured in Hertz. The other feature which is constant with time at any given frequency is δ, the phase difference in radians between the peak value of the stress and the peak value of the strain. These two values, G^* and δ, are characteristic of the material. It is straightforward for us to visualise the situation where an elastic solid is placed between the cone and plate. When a tangential displacement is applied to the lower plate a strain in the sample is produced. That displacement is transmitted directly through the sample. The upper cone will react in proportion to the applied strain to give a stress response. An oscillating strain will give an oscillating stress response which is *in phase* with the strain so δ will be zero. However, if we have a Newtonian liquid in our instrument, the peak stress is out of phase by $\pi/2$ rad because the peak stress is proportional to the *rate of strain*. So it follows that if we have a viscoelastic material in our instrument, so that some energy is stored and some dissipated, the stored contribution will be in phase whilst the dissipated or loss contribution will be out of phase with the applied strain.

In order to describe the material properties as a function of frequency for a body that behaves as a Maxwell model we need to use the constitutive equation. This is given in Equation (4.8), which describes the relationship between the stress and the strain. It is most convenient to express the applied sinusoidal wave in the exponential form of complex number notation:

$$\gamma^* = \gamma_0 \exp(i\omega t); \quad \dot{\gamma}^* = i\omega\gamma_0 \exp(i\omega t) = i\omega\gamma^* \tag{4.19}$$

Now the stress response is shifted through a phase angle δ:

$$\sigma^* = \sigma_0 \exp[i(\omega t + \delta)]; \quad \dot{\sigma}^* = i\omega\sigma_0 \exp[i(\omega t + \delta)] = i\omega\sigma^* \tag{4.20}$$

So if we substitute the complex stress and strains into the constitutive equation for a Maxwell fluid the resulting relationship is given by Equation (4.21):

$$\dot{\gamma}^* = \frac{\dot{\sigma}^*}{G} + \frac{\sigma^*}{\eta} \tag{4.21}$$

and using Equations (4.19) and (4.20):

$$i\omega\gamma^* = \frac{i\omega\sigma*}{G} + \frac{\sigma*}{\eta} \tag{4.22}$$

or rearranging we have:

$$\frac{\gamma^*}{G\sigma^*} = 1 + \frac{G}{i\omega\eta} \tag{4.23}$$

The ratio of the stress to the strain is the complex modulus $G^*(\omega)$. We can rearrange this expression to give the complex modulus and the frequency, and using Equation (4.14) we have:

$$\frac{G}{G^*(\omega)} = 1 + \frac{1}{i\omega\tau_m} \tag{4.24}$$

This can be further rearranged to give an expression in terms of $G^*(\omega)$:

$$G^*(\omega) = G\left(\frac{i\omega\tau_m}{1 + i\omega\tau_m}\right) \tag{4.25}$$

This expression describes the variation of the complex modulus with frequency for a Maxwell model. It is normal to separate the real and imaginary components of this expression. This is achieved by multiplying through by $(1 - i\omega\tau)$ to give

$$G^*(\omega) = G'(\omega) + iG''(\omega) = G\frac{(\omega\tau_m)^2}{1 + (\omega\tau_m)^2} + iG\frac{\omega\tau_m}{1 + (\omega\tau_m)^2} \tag{4.26}$$

where $G'(\omega)$ is called the storage modulus and $G''(\omega)$ the loss modulus:

$$G'(\omega) = G\frac{(\omega\tau_m)^2}{1 + (\omega\tau_m)^2} \tag{4.27}$$

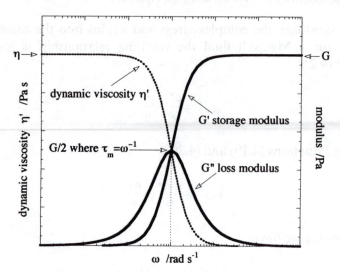

Figure 4.7   *The frequency response of a Maxwell model*

$$G''(\omega) = G \frac{\omega\tau_m}{1 + (\omega\tau_m)^2} \tag{4.28}$$

Now these expressions describe the frequency dependence of the stress with respect to the strain. It is normal to represent these as two moduli which determine the component of stress in phase with the applied strain (storage modulus) and the component out of phase by 90°. The functions have some identifying features. As the frequency increases, the loss modulus at first increases from zero to $G/2$ and then reduces to zero giving the bell-shaped curve in Figure 4.7. The maximum in the curve and crossover point between storage and loss moduli occurs at τ_m.

The storage modulus increases from zero to a maximum value; in fact as the frequency tends to infinity the complex modulus and $G'(\omega)$ equate, *i.e.*

$$G^*(\omega \to \infty) = G'(\omega \to \infty) = G \tag{4.29}$$

The result that the high frequency limit of the storage modulus equates with the elastic modulus of the spring means that for a single Maxwell it is possible to replace G by $G(\infty)$. Conceptually the storage modulus is related to the energy stored in the elastic contributions within the sample. By integrating the stress during one cyclic period of the strain we can calculate the average energy stored in the material per unit volume in one cycle:

$$E^s = \frac{G'(\omega)\gamma_0^2}{4} \tag{4.30}$$

and the energy dissipated per unit volume is

$$E^l = \pi G''(\omega)\gamma_0^2 \tag{4.31}$$

These expressions are general for any viscoelastic fluid in the linear region. For a Maxwell model the ratio of these values is given by Equation (4.32):

$$\frac{E^l}{E^s} = \frac{4\pi}{\omega\tau_m} \tag{4.32}$$

You can see that until $\omega\tau_m$ is greater than 4π more energy is dissipated than stored. In other words the curves in Figure 4.7 do not directly represent the energy stored as a function of frequency and it is incorrect to say that at the crossover point the rate of energy dissipation and storage per unit volume is the same. The storage and loss moduli are subtle descriptions of the material properties of the system. These two properties are related to the phase angle and complex modulus. These are both functions of the applied frequency and represent an alternative description of the system:

$$G'(\omega) = G^*(\omega)\cos\delta \tag{4.33}$$

$$G'(\omega) = G^*(\omega)\sin\delta \tag{4.34}$$

The phase angle changes with frequency and this is shown in Figure 4.7. As the frequency increases the sample becomes more elastic. Thus the phase difference between the stress and the strain reduces. There is an important feature that we can obtain from the dynamic response of a viscoelastic model and that is the dynamic viscosity. In oscillatory flow there is an analogue to the viscosity measured in continuous shear flow. We can illustrate this by considering the relationship between the stress and the strain. This defines the complex modulus:

$$G^*(\omega) = \frac{\sigma^*}{\gamma^*} = G'(\omega) + iG''(\omega) \tag{4.35}$$

Now if we use the relationship in Equation (4.19) relating the strain and rate of strain we get

$$\sigma^* = \left[\frac{G'(\omega) + iG''(\omega)}{i\omega}\right]\dot{\gamma}^* \qquad (4.36)$$

where the term in the square brackets is analogous to viscosity because it relates shear stress to shear rate. We can express this as a complex or dynamic viscosity $\eta^*(\omega)$ (and noting that $i^{-1} = -1$):

$$\eta^*(\omega) = \eta'(\omega) - i\eta''(\omega) \qquad (4.37)$$

The term in phase with the rate of strain is related to the loss modulus and the term out of phase is related to the storage modulus:

$$\eta'(\omega) = \frac{G''(\omega)}{\omega} = \frac{G\tau}{1 + (\omega\tau)^2}; \quad \eta''(\omega) = \frac{G'(\omega)}{\omega} = \frac{G\omega\tau^2}{1 + (\omega\tau)^2} \qquad (4.38)$$

The functions have some important features. As the frequency reduces the value of $\eta'(\omega)$ increases from zero to a maximum value. In fact as the frequency tends to zero the complex viscosity and $\eta'(\omega)$ equate, *i.e.*

$$\eta^*(\omega \to 0) = \eta'(\omega \to 0) = \eta \qquad (4.39)$$

This result, that the low frequency limit of the in phase component of the viscosity equates to the viscosity of the dashpot, means that for a single Maxwell model it is possible to replace η by $\eta(0)$. Thus far we have concentrated on the description of experimental responses to the application of a strain. Similar constructions can be developed for the application of a stress. For example the application of an oscillating stress to a sample gives rise to an oscillating strain. We can define a complex compliance J^* which is the ratio of the strain to the stress. We will explore the relationship between different experiments and the resulting models in Section 4.6.

4.4.4 Multiple Processes

A stress relaxation experiment can be performed on a wide range of materials. If we perform such a test on a real material a number of deviations are normally observed from the behaviour of a single Maxwell model. Some of these deviations are associated with the application of the strain itself. For example it is very difficult to apply an instantaneous strain to a sample. This influences the measured response at short experimental times. It is often difficult to apply a strain small enough to provide a linear response. A Maxwell model is only applicable to linear responses. Even if you were to imagine an experiment where a strain is instantaneously applied and the sample responds as a linear viscoelastic material the relaxation need not follow a single Maxwell model. There are two main approaches to dealing with this problem. The approach we will adopt in this and the following section is to assume that the material can be described by a combination of springs and dashpots, added together to give an overall material response, but where the relaxation of any spring is not affected by the response of the others. In Section 4.6 we will consider what happens if the material cannot be described by a simple addition of springs and dashpots. Let us begin by briefly considering the origin of the relaxation time in the Maxwell model. We have seen (Section 4.2) that the characteristic diffusional time for a system of spherical particles depends upon the cube of the particle radius. This represents the characteristic distance moved by the particle in time τ which would cause the relaxation of the stress from the applied strain. However, most molecules are far from spherical and there are many systems that are mixtures of different molecules. For many systems the idea that there is just a single relaxation time is too simple. At any instant in time the thermal motion of the molecules or particles in the system imply a distribution of interparticle spaces and particle–particle interactions. This gives a range of relaxation times with accompanying viscous and elastic contributions. One way in which we can model this distribution of molecular interactions with their characteristic relaxation times is to imagine each is replaced by a single Maxwell model. The overall relaxation is formed by summing all these processes together. A model of this is shown schematically in Figure 4.8 and is called a multiple or generalised Maxwell model.

Suppose the multiple Maxwell model which describes the material we are interested in is composed of m processes each with an elasticity G_j, a viscous process with a viscosity η_j and a corresponding relaxation time τ_j. We can form the relaxation function by adding all these models together:

$$G(t) = \sum_{j=1}^{m} G_j \exp\left(-\frac{t}{\tau_j}\right) \qquad (4.40)$$

Now we can visualise that each of these models represents a relaxation process in the system. However, it is likely that some spacings, or sizes, and interactions occur more often than others, so that each process is not unique. We could represent the m models by a series of unique models where each model occurs p_i times. There would be n such models where n is less than m:

$$G(t) = \sum_{i=1}^{n} p_i G_i \exp\left(-\frac{t}{\tau_i}\right) \qquad (4.41)$$

If we divide the probability of occurrence by the number of models m we obtain the frequency of occurrence f_i:

$$G(t) = \sum_{i=1}^{n} f_i (mG_i) \exp\left(-\frac{t}{\tau_i}\right) \qquad (4.42)$$

The product of m and G_j could be replaced by a 'new' elasticity, which for convenience we could write as G_j. Equations (4.41) and (4.42) describe the relaxation of the stress in terms of a distribution of discrete processes. Some processes may be broadly separated in time to give two clearly separated relaxation processes. Other systems may show a distribution of elastic processes very closely separated in time.

One feature of the Maxwell model is that it allows the complete relaxation of any applied strain, *i.e.* we do not observe any energy stored in the sample, and all the energy stored in the springs is dissipated in flow. Such a material is termed a viscoelastic fluid or viscoelastic liquid. However, it is feasible for a material to show an apparent yield stress at low shear rates or stresses (Section 6.2). We can think of this as an elastic response at low stresses or strains regardless of the application time (over all practical timescales). We can only obtain such a response by removing one of the dashpots from the viscoelastic model in Figure 4.8. When a

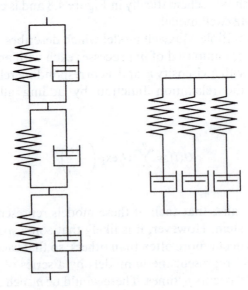

Figure 4.8   *A multiple Maxwell model and a multiple Kelvin–Voigt model*

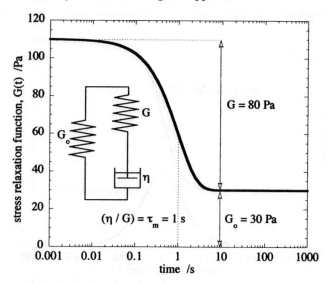

Figure 4.9 *A stress relaxation curve for a viscoelastic solid*

strain is applied to this model all the energy stored in the springs with an accompanying dashpot is dissipated in the flow. However, the spring in isolation has no mechanism by which it can relax the applied strain. Elastic energy will always be stored in this spring. Thus at long times after all the other spring and dashpot pairs have relaxed, the stress will be maintained. The stress relaxation function now appears as in Figure 4.9, with the relaxation function displaced from the baseline by the value of the spring G_0, which is equal to the low frequency limit in oscillation, $G(0)$.

This material is a linear viscoelastic solid and is described by the multiple Maxwell model with an additional term, the spring elasticity $G(0)$:

$$G(t) = \sum_{i=1}^{n} p_i G_i \exp\left(-\frac{t}{\tau_i}\right) + G(0) \tag{4.43}$$

These same ideas apply equally well to the application of an oscillating stress or strain. For a viscoelastic liquid we obtain the following results:

$$G'(\omega) = \sum_{i=1}^{n} p_i G_i \frac{(\omega\tau)^2}{1 + (\omega\tau)^2} \tag{4.44}$$

$$G''(\omega) = \sum_{i=1}^{n} p_i G_i \frac{\omega\tau}{1 + (\omega\tau)^2} \tag{4.45}$$

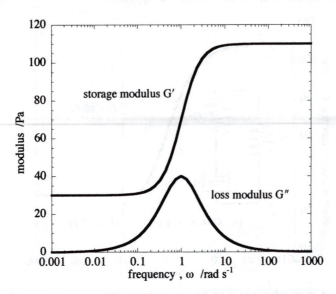

Figure 4.10   *The storage and loss moduli for the viscoelastic solid in Figure 4.9*

For a viscoelastic solid, the loss modulus which reflects the viscous processes in the material is unaffected by the presence of a spring without a dashpot. The storage modulus includes the elastic component $G(0)$:

$$G''(\omega) = \sum_{i=1}^{n} p_i G_i \frac{(\omega\tau)^2}{1 + (\omega\tau)^2} + G(0) \tag{4.46}$$

This results in an elastic contribution at all frequencies. This is shown in Figure 4.10 where the storage modulus is displaced from the baseline by $G(0)$ at low frequencies.

4.4.5   A Spectral Approach To Linear Viscoelastic Theory

We have developed the idea that we can describe linear viscoelastic materials by a sum of Maxwell models. These models are the most appropriate for describing the response of a body to an applied strain. The same ideas apply to a sum of Kelvin models, which are more appropriately applied to stress controlled experiments. A combination of these models enables us to predict the results of different experiments. If we were able to predict the form of the model from the chemical constituents of the system we could predict all the viscoelastic responses in shear. We know that when a strain is applied to a viscoelastic material the molecules and particles that form the system gradual diffuse to relax the applied strain. For example, consider a solution of polymer molecules. If we were to consider a single polymer molecule in solution we can visualise the strain both moving the molecule and displacing the segments forming the molecule. The strain is relaxed by the diffusion of the segments eventually achieving the natural conformation of the polymer. We could consider each process to have a characteristic time-scale of relaxation and associated viscous drag. Each mode of relaxation would have an associated Maxwell model. If we were to sum all these models together we could construct a multiple Maxwell model appropriate to that polymer chain. We must recognise that whilst these ideas give rise to a model for the system, a typical system is far more complex. In reality a polymerisation produces a continuous distribution of weights rather than a single molecular weight. There must therefore exist a corresponding continuous distribution of relaxation processes. This can be reflected by using a relaxation mechanism consistent with that shown by a Maxwell model but replacing a discrete distribution of behaviour with a continuous distribution. A continuous distribution of mechanisms relaxing as a Maxwell model is termed a *relaxation spectrum H*. The corresponding distribution in stress for a Kelvin model is termed a *retardation spectrum L*. In order to evaluate the relaxation function the sum needs to be replaced by an integral over all such processes. If we examine the response of a single Maxwell model in Figure 4.7 we can see it gives a variation in loss modulus over four orders of magnitude in frequency. The data is plotted on a log scale to illustrate this. Rheological relaxation processes tend to occur over very long timescales and so a relaxation spectrum is most appropriately modelled in logarithmic time. The limits of the integral must encompass the slowest process at time $\tau \to 0$ and the fastest at time $\tau \to \infty$. On a logarithmic timescale this corresponds to $\tau = -\infty$ to $\tau = +\infty$. The relaxation function for a viscoelastic liquid is described by:

$$G(t) = \int_{-\infty}^{+\infty} H \exp(-t/\tau) \mathrm{d} \ln \tau \tag{4.47}$$

The relaxation spectrum H is independent of the experimental time t and is a fundamental description of the system. The exponential function depends upon both the experimental time and the relaxation time. Such a function in the context of this integral is called the kernel. In order to describe different experiments in terms of a relaxation spectrum H or retardation spectrum L it is the kernel that changes. The integral can be formed in time or frequency depending upon the experiment being modelled. The inclusion of elastic properties at all frequencies and times can be achieved by including an additional process in the relaxation spectrum at infinite time or more simply by adding the value of $G(0)$ to the integral:

$$G(t) = G(0) + \int_{-\infty}^{+\infty} H \exp(-t/\tau) d\ln\tau \tag{4.48}$$

The relaxation spectrum greatly influences the behaviour observed in experiments. As an example of this we can consider how the relaxation spectrum affects the storage and loss moduli. To evaluate this we need to change the kernel to that for a Maxwell model in oscillation and replace the experimental time by oscillation frequency:

$$G'(\omega) = \int_{-\infty}^{+\infty} H \frac{(\omega\tau)^2}{1+(\omega\tau)^2} d\ln\tau \tag{4.49}$$

$$G''(\omega) = \int_{-\infty}^{+\infty} H \frac{\omega\tau}{1+(\omega\tau)^2} d\ln\tau \tag{4.50}$$

$$\eta'(\omega) = \int_{-\infty}^{+\infty} H \frac{\tau}{1+(\omega\tau)^2} d\ln\tau \tag{4.51}$$

$$\eta''(\omega) = \int_{-\infty}^{+\infty} H \frac{\omega\tau^2}{1+(\omega\tau)^2} d\ln\tau \tag{4.52}$$

In the limit of high frequencies the integral for the loss modulus tends to zero as the denominator in Equation 4.50 tends to infinity. The storage modulus tends to $G(\infty)$ which is just the integral under the relaxation spectrum:

$$G(\infty) = \int_{-\infty}^{+\infty} H d\ln\tau \tag{4.53}$$

In the limit of low frequencies the integral for the loss in the viscosity tends to zero. The storage term tends to $\eta(0)$ which is the integral under the relaxation spectrum after it has been multiplied by the appropriate τ value at each point:

$$\eta(0) = \int_{-\infty}^{+\infty} H\tau d\ln\tau \tag{4.54}$$

We can illustrate the effect of the relaxation spectrum by assuming a

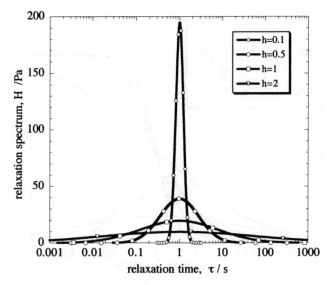

Figure 4.11 *A log normal relaxation spectrum centred on a modal time of 1 s*

form for the distribution. A log normal distribution will illustrate the effect on the storage and loss moduli:

$$H = \frac{G(\infty)}{\sqrt{8.405\pi h^2}} \exp\left(\frac{-[\ln(\tau/\tau_0)]^2}{8.405h^2}\right) \tag{4.55}$$

The spectrum has been selected such that the integral under the distribution gives $G(\infty)$. The distribution is centred on the relaxation time τ_0 and the width of the peak is given by h. The term h is the half-width of the peak at half of the maximum height. A range of distributions with different half width half heights is shown in Figure 4.11. Now we can calculate the storage and loss moduli for a material that is described by these distributions. This is shown in Figure 4.12.

The range of frequencies used to calculate the moduli are typically available on many instruments. The important feature that these calculations illustrate is that as the breadth of the distributions is increased the original sigmoidal and bell shaped curves of the Maxwell model are progressively lost. A distribution of Maxwell models can produce a wide range of experimental behaviour depending upon the relaxation times and the elastic responses present in the material. The relaxation spectrum can be composed of more than one peak or could contain a simple Maxwell process represented by a spike in the distribution. This results in complex forms for all the elastic moduli.

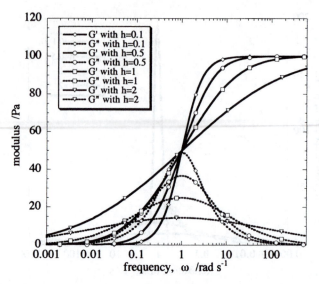

Figure 4.12 *The storage and loss moduli corresponding to the relaxation spectra in Figure 4.11*

4.5 LINEAR VISCOELASTIC EXPERIMENTS

There are many types of deformation and forces that can be applied to material. One of the foundations of viscoelastic theory is the Boltzmann Superposition Principle. This principle is based on the assumption that the effects of a series of applied stresses acting on a sample results in a strain which is related to the sum of the stresses. The same argument applies to the application of a strain. For example we could apply an instantaneous stress to a body and maintain that stress constant. For a viscoelastic material the strain will increase with time. The ratio of the strain to the stress defines the compliance of the body:

$$J(t) = \frac{\gamma(t)}{\sigma} \qquad (4.56)$$

Now suppose at a time t_1 later we apply another stress σ_1. The strain will respond to this additional stress. This can be predicted by adding the result of this additional stress on the strain at a time $t - t_1$ later:

$$\gamma(t) = \sigma J(t) + \sigma_1 J(t - t_1) \qquad (4.57)$$

It should be remembered that $\sigma_1 = 0$ while $t < t_1$. This linear superposition of stresses can be generalised to any number of applied stresses:

$$\gamma(t) = \sum_{t_i=-\infty}^{t_i=t} \sigma_i J(t - t_i) \qquad (4.58)$$

The sum in this expression can be replaced by an integral which will enable us to describe the strain response to any stress history:

$$\gamma(t) = \int_{-\infty}^{t} J(t - t')\dot{\sigma}(t')\mathrm{d}t' \qquad (4.59)$$

There is an analogous relationship for a series of strains:

$$\sigma(t) = \int_{-\infty}^{t} G(t - t')\dot{\gamma}(t')\mathrm{d}t' \qquad (4.60)$$

These two mathematical Equations (4.59) and (4.60) illustrate an important feature about linear viscoelastic measurements, *i.e.* the central role played by the relaxation function and the compliance. These terms can be used to describe the response of a material to any deformation history. If these can be modelled in terms of the chemistry of the system the complete linear rheological response of our material can be obtained.

Finally it is worth noting an alternate form for the stress dependence of a series of strains. Some microstructural models utilise the memory function $m(t)$. This is the rate of change of the stress relaxation function:

$$m(t) = -\frac{\mathrm{d}G(t)}{\mathrm{d}t} \qquad (4.61)$$

The superposition integral becomes, for a viscoelastic liquid,

$$\sigma(t) = \int_{-\infty}^{t} m(t - t')\gamma(t, t')\mathrm{d}t \qquad (4.62)$$

The term $\gamma(t,t')$ is the shear strain at time t' relative to the strain at time t. The use of a memory function has been adopted in polymer modelling. For example this approach is used by Doi and Edwards[11] to describe linear responses of solution polymers which they extended to non-linear viscoelastic responses in both shear and extension.

4.5.1 Relaxation

The ideal stress relaxation experiment is one in which the stress is instantaneously applied. We have seen in Section 4.4.2 the exponential relaxation that characterises the response of a Maxwell model. We can consider this experiment in detail as an example of the application of the Boltzmann Superposition Principle. The practical application of an instantaneous strain is very difficult to achieve. In a laboratory experi-

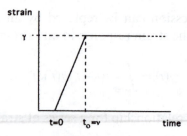

Figure 4.13 *The applied strain in a typical stress relaxation experiment*

ment the strain rises to its maximum value over a finite time. We can represent this feature as shown in Figure 4.13.

Let us suppose the strain applied at time t_0 increases over a time v to a maximum value γ. At times less than $t_0 - v$ no strain is applied and at times greater than t_0 the strain is constant. This gives the limits to the Boltzmann superposition integral:

$$\dot{\gamma} = 0 \quad t < t_0 - v$$
$$\dot{\gamma} = \gamma/v \quad t_0 - v \leq t \leq t_0$$
$$\dot{\gamma} = 0 \quad t > t_0$$

Now we can apply these conditions to the Boltzmann superposition integral (Equation 4.60):

$$\sigma(t) = \int_{t_0-v}^{t_0} G(t - t')\left(\frac{\gamma}{v}\right)\mathrm{d}t' \tag{4.63}$$

In principle this integral could be applied directly to the Maxwell model to predict the decay of stress at any point in time. We can simplify this further with an additional assumption that is experimentally verified, *i.e.* that the function in the integral is continuous. The first value for the mean theorem for integrals states that if a function f(x) is continuous between the limits a and b there exists a value f(q) such that

$$\int_a^b \mathrm{f}(x)\mathrm{d}x = \mathrm{f}(q)(b - a) \tag{4.64}$$

where q lies between a and b. We can apply this expression to the superposition integral to give

$$\sigma(t) = G(q)\left(\frac{\gamma}{v}\right)[t_0 - (t_0 - v)] \tag{4.65}$$

and so we get from this expression

$$\sigma(t) = \gamma G(q) \tag{4.66}$$

Now we know that the value of q lies between the limits $t_0 - v$ and t_0. We can choose time $t_0 = 0$. We do not have a value for q but we know that it occurs at $t + gv$ where g lies between 0 and 1:

$$\sigma(t) = \gamma G(t + gv) \tag{4.67}$$

Providing $gv \ll t$ then for a viscoelastic liquid we obtain

$$\sigma(t) = \gamma G(t) \tag{4.68}$$

In an experiment when we apply a step strain the rate of application of the strain influences the relaxation of the stress at short times. There are other factors which can influence the response that is observed. For example it is common for elastic samples to resonate with the applied actuator and transmit transient waves through the sample. This can lead to fluctuations in the stress at short times. A typical example is shown in Figure 4.14.

It is difficult to predict these responses directly because they depend upon a wide range of instrumental properties in addition to the material properties of the sample. The onset of this behaviour can be explored through the use of a stress growth experiment.

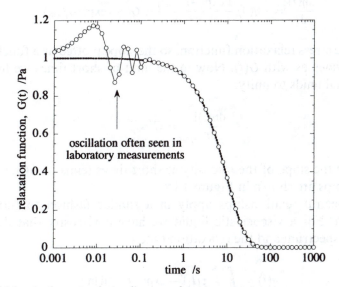

Figure 4.14 *A schematic of a 'real' stress relaxation response*

4.5.2 Stress Growth

The application of a linearly ramped strain can provide information on both the sample elasticity and viscosity. The stress will grow in proportion to the applied strain. The ratio of the strain over the applied time gives the shear rate. Applying the Boltzmann Superposition Principle we obtain the following expression:

$$\sigma(t) = \dot{\gamma} \int_0^t G(t - u)\mathrm{d}u \tag{4.69}$$

Now we can substitute a Maxwell model for the relaxation function:

$$\frac{\sigma(t)}{\dot{\gamma}} = \eta(t) = G(\infty) \int_0^t \exp - \frac{(t - u)}{\tau_m}\mathrm{d}u$$

so that

$$\eta(t) = \frac{\sigma(t)}{\dot{\gamma}} = \eta(0)[1 - - \exp(-t/\tau_m)] \tag{4.70}$$

Now when we apply a shear rate to the sample, as $t \to \infty$ the viscosity tends to $\eta(0)$. At short times the elasticity can be obtained by differentiating the viscosity versus time curve:

$$\frac{\mathrm{d}\eta(t)}{\mathrm{d}t} = -\eta(0)\frac{\mathrm{d}[\exp(-t/\tau_m)]}{\mathrm{d}t} = G(\infty)\exp(-t/\tau_m) \tag{4.71}$$

This is the stress relaxation function, so the slope plotted as a function of time provides us with $G(t)$. Now in the limit of short times we find the exponential tends to unity:

$$\left.\frac{\mathrm{d}\eta(t)}{\mathrm{d}t}\right|_{t\to 0} = G(\infty) \tag{4.72}$$

Therefore the slope of the viscosity at short times tends to $G(\infty)$. These relationships are shown in Figure 4.15.

The spectral relationships apply in a similar fashion to those for relaxation. For a viscoelastic liquid we have a viscosity that depends upon the spectrum and the relaxation time τ:

$$\eta(t) = \int_{-\infty}^{+\infty} \tau H[1 - \exp(-t/\tau)]\mathrm{d}\ln \tau \tag{4.73}$$

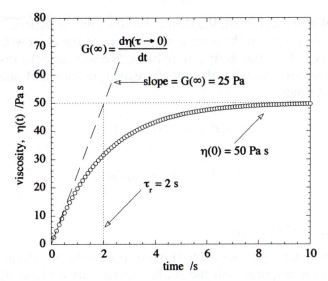

Figure 4.15 *The stress growth function for a Maxwell model with a relaxation time* τ_r

For a viscoelastic solid the situation is more complex because the solid component will never flow. As the strain is applied with time the stress will increase continually with time. The sample will show no plateau viscosity, although there may be a low shear viscous contribution. This applies to both a single Maxwell model and one with a spectrum of processes:

$$\eta(t) = \eta(0)[1 - \exp(-t/\tau)] + G(0)t \tag{4.74}$$

$$\eta(t) = \int_{-\infty}^{+\infty} \tau H[1 - \exp(-t/\tau)]\mathrm{d}\ln\tau + G(0)t \tag{4.75}$$

In both cases the apparent viscosity increases with time.

4.5.3 Anti-thixotropic Response

An important and sometimes overlooked feature of all linear viscoelastic liquids that follow a Maxwell response is that they exhibit anti-thixotropic behaviour. That is if a constant shear rate is applied to a material that behaves as a Maxwell model the viscosity increases with time up to a constant value. We have seen in the previous examples that as the shear rate is applied the stress progressively increases to a maximum value. The approach we should adopt is to use the Boltzmann Superposition Principle. Initially we apply a continuous shear rate until a steady state viscosity of $\eta(0)$ is achieved. We then cease to strain the material and the stress decays to zero in the inverse manner of the stress growth. Hence the viscosity falls as the shear rate reduces. We will state the results for both a viscoelastic liquid with a single relaxation time and one with a spectrum of times:

$$\eta(t) = \eta(0)\exp(-t/\tau) \tag{4.76}$$

$$\eta(t) = \int_{-\infty}^{+\infty} \tau H \exp(-t/\tau)\mathrm{d}\ln\tau \tag{4.77}$$

4.5.4   Creep and Recovery

So far we have largely concentrated on the application of a strain and the resulting stress response. All the strain experiments we have discussed thus far have analogues in the stress regime. Perhaps the most important of these are the creep and recovery functions. Here we apply a step stress to the sample, σ_0, and we achieve a strain response. As with the stress relaxation experiment where it is difficult to apply an instantaneous strain, it is equally difficult to apply an instantaneous stress. In addition the instantaneous strain response is prone to oscillations at short times. The Boltzmann Superposition Principle can be applied and we obtain the result shown in Equation (4.56) for the compliance. The compliance is sometimes referred to as the creep compliance. After the stress has been applied for a time t_1 it is removed. If there has been no flow during the application of the stress the strain is gradually recovered and falls to zero. If there has been some flow the recovery will be incomplete. We have two zones of physical responses, the creep region $\gamma_c(t)$ and the recovery region $\gamma_r(t)$:

$$\gamma_c = \sigma_0 J(t) \tag{4.78}$$

$$\gamma_r(t) = \sigma_0[J(t1) - J(t - t_1)] \tag{4.79}$$

In order to obtain a general model of the creep and recovery functions we need to use a Kelvin model or a Kelvin kernel and retardation spectrum L. However, there are some additional subtleties that need to be accounted for. One of the features of a Maxwell model is that it possesses a high frequency limit to the shear modulus. This means there is an instantaneous response at all strains. The response of a simple Kelvin model is shown in Equation 4.80:

$$J_1(t) = \frac{[1 - \exp(-t/\tau_k)]}{G} = J[1 - \exp(-t/\tau_k)] \tag{4.80}$$

Here the term τ_k is the retardation time. It is given by the product of the compliance of the spring and the viscosity of the dashpot. If we examine this function we see that as $t \to 0$ the compliance tends to zero and hence the elastic modulus tends to infinity. Whilst it is philosophically possible to simulate a material with an infinite elastic modulus, for most situations it is not a realistic model. We must conclude that we need an additional term in a single Kelvin model to represent a 'typical' material. We can achieve this by connecting an additional spring in series to our model with a compliance J_g. This is known from the polymer literature as the 'standard linear solid' and J_g is the 'glassy compliance':

$$\gamma_c(t) = \sigma_0 \big[J_g + J_1(t) \big] \tag{4.81}$$

$$\gamma_r(t) = \sigma_0 \big\{ J_g + J_1(t_1) - [J_1(t - t_1) + J_g] \big\} \tag{4.82}$$

The response of this body is shown in Figure 4.16.

We can see that as the stress is applied the strain increases up to a time $t = t_1$. Once the stress is removed we see complete recovery of the strain. All the strain stored has been recovered. The material has the properties of an elastic solid. In order to achieve viscous flow we need to include an additional term, the viscous loss term. This is known as a 'Burger Body'

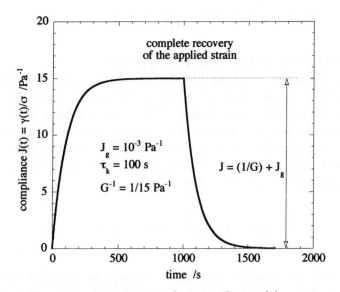

Figure 4.16 *The creep and recovery curve for a viscoelastic solid*

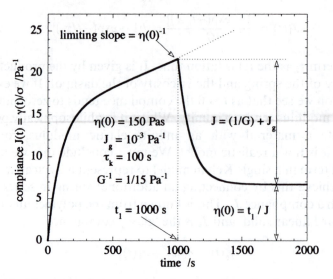

Figure 4.17   *The creep and recovery curve for a viscoelastic liquid*

and a dashpot, with a viscosity equivalent to $\eta(0)$, is added in series to a standard linear solid:

$$\gamma_c(t) = \sigma_0 \left(J_g + J_1(t) + \frac{t}{\eta(0)} \right) \qquad (4.83)$$

$$\gamma_r(t) = \sigma_0 \left(J_g + J_1(t_1) + \frac{t_1}{\eta(0)} - \left[J_1(t - t_1) + J_g \right] \right) \qquad (4.84)$$

The response of this body is shown in Figure 4.17.

As with the elastic solid we can see that as the stress is applied the strain increases up to a time $t = t_1$. Once the stress is removed we see partial recovery of the strain. Some of the strain has been dissipated in viscous flow. Laboratory measurements often show a high frequency oscillation at short times after a stress is applied or removed just as is observed with the stress relaxation experiment. We can replace a Kelvin model by a distribution of retardation times:

$$J_1(t) = \int_{-\infty}^{+\infty} L[1 - \exp(-t/\tau)] \mathrm{d} \ln \tau \qquad (4.85)$$

Now as $t \to 0$ the spectral function L reduces to the area under the distribution. This is the steady state compliance J_e:

$$J_e = \int_{-\infty}^{+\infty} L \, d\ln\tau \qquad (4.86)$$

The contribution of this component is shown in Figures 4.16 and 4.17. One of the important features to recognise about the retardation spectrum is that it only has an indirect relationship to both the zero shear rate viscosity and the high frequency shear modulus. Both these properties are contained in the relaxation spectra. We shall see in Section 4.5.7 that, whilst a relationship exists between H and L it is somewhat complex.

4.5.5 Strain Oscillation

So far we have seen that if we begin with the Boltzmann superposition integral and include in that expression a mathematical representation for the stress or strain we apply, it is possible to derive a relationship between the instrumental response and the properties of the material. For an oscillating strain the problem can be solved either using complex number theory or simple trigonometric functions for the deformation applied. Suppose we apply a strain described by a sine wave:

$$\gamma = \gamma_0 \sin(\omega t) \qquad (4.87)$$

Now in order to apply the Boltzmann Superposition Principle (Equation 4.60) we need to express this as a strain rate. Differentiating with respect to time gives us

$$\dot{\gamma} = \omega\gamma_0 \cos(\omega t) \qquad (4.88)$$

Now the Boltzmann superposition integral is given by Equation (4.60). Substituting for the strain and replacing t by t' in Equation (4.88) gives

$$\sigma(t) = \omega\gamma_0 \int_{-\infty}^{t} G(t - t') \cos(\omega t') dt \qquad (4.89)$$

For convenience we can change variables and the integral limits so that $s = t - t'$:

$$\sigma(t) = \omega\gamma_0 \int_{0}^{\infty} G(s) \cos[\omega(t - s)] ds \qquad (4.90)$$

Simple trigonometry allows the cosine term to be rearranged:

$$\sigma(t) = \gamma_0 \sin(\omega t)\left(\omega \int_0^\infty G(s)\sin(\omega s)ds\right) + \gamma_0 \cos(\omega t)\left(\omega \int_0^\infty G(s)\cos(\omega s)ds\right)$$

$$(4.91)$$

Now the first term on the right is in phase with the applied strain, *i.e.* it has the form of a sine wave. This can be equated with the storage modulus. Conversely the phase difference between the second term on the right and the applied signal is the difference between sine and cosine waves which can be equated with the loss modulus:

$$G'(\omega) = \omega \int_0^\infty G(s)\sin(\omega s)ds \qquad (4.92)$$

$$G''(\omega) = \omega \int_0^\infty G(s)\cos(\omega s)ds \qquad (4.93)$$

These are one-sided Fourier transforms, which will be discussed further in Section 4.6.3. With these definitions we can substitute back into the above expression to give

$$\sigma(t) = \gamma_0 \sin(\omega t)G'(\omega) + \gamma_0 \cos(\omega t)G''(\omega) \qquad (4.94)$$

Now as we have stated earlier we can represent the phase difference between the applied stress and strain as δ. So for a peak stress σ_0 we can visualise a stress displaced by a phase difference δ:

$$\sigma(t) = \sigma_0 \sin(\omega t + \delta) \qquad (4.95)$$

Trigonometry gives us

$$\sigma(t) = \sigma_0 \cos(\delta)\sin(\omega t) + \sigma_0 \sin(\delta)\cos(\omega t) \qquad (4.96)$$

Thus equating (4.94) and (4.96) we obtain expressions for the storage and loss modulus:

$$G'(\omega) = \frac{\sigma_0}{\gamma_0}\cos(\delta) = G^*(\omega)\cos(\delta) \qquad (4.97)$$

$$G''(\omega) = \frac{\sigma_0}{\gamma_0}\sin(\delta) = G^*(\omega)\sin(\delta) \qquad (4.98)$$

which is the result from Equations (4.33) and (4.44). In complex number form we obtain

$$G^* = G'(\omega) + iG''(\omega) \qquad (4.99)$$

4.5.6 Stress Oscillation

The application of an oscillating stress gives a corresponding response in strain. The Boltzmann Superposition Principle may again be utilised as our starting point. The creep function is used rather than the stress relaxation function. The derivation follows much the same course as that for a strain lead oscillation, returning similar results for the complex compliance. The storage compliance $J'(\omega)$ and loss compliance $J''(\omega)$ represent contributions to the energy stored and lost in the system, although their relationship to these parameters is not as straightforward as given by Equations (4.30) and (4.31). They contain both the viscosity and the elasticity:

$$J^*(i\omega) = [J_g + J'(\omega)] + i\left(-\frac{1}{\omega\eta(0)} + J''(\omega)\right) \qquad (4.100)$$

The corresponding compliance functions are

$$J'(\omega) = \omega \int_0^\infty J(t)\sin(\omega t)dt \qquad (4.101a)$$

$$J''(\omega) = -\omega \int_0^\infty J(t)\cos(\omega t)dt \qquad (4.101b)$$

In the same manner as the modulus can be related to the relaxation spectrum so the compliance can be related to the retardation spectrum:

$$J'(\omega) = \int_{-\infty}^{+\infty} L\frac{1}{1+\omega^2\tau^2}d\ln\tau \qquad (4.102)$$

$$J''(\omega) = \int_{-\infty}^{+\infty} L\frac{\omega\tau}{1+\omega^2\tau^2}d\ln\tau \qquad (4.103)$$

In the limit of zero frequency we obtain

$$J_e = J'(\omega \to 0) = \int_{-\infty}^{+\infty} L d\ln\tau J_e = J'(\omega \to 0) = \int_{-\infty}^{+\infty} L d\ln\tau \qquad (4.104)$$

which is the steady state compliance.

4.6 INTERRELATIONSHIPS BETWEEN THE MEASUREMENTS AND THE SPECTRA

The mathematics which underpins the theory of linear viscoelasticity is common to many simple time varying processes. It has already been indicated that linear viscoelastic experiments are a form of mechanical spectroscopy. The closest analogue to linear viscoelasticity is that of direct and alternating current applied to electrical circuits of resistors and capacitors. The resistor, which dissipates energy, is an analogue of a dashpot and the capacitor, storing energy, is the analogue of a spring. The first impression is that this behaviour – whilst it might be analogous mathematically to rheological processes – has little to do with the chemistry of systems. However one should recall that conductivity measurements made on ac bridges are routinely performed to determine the purity of water relative to its ion content or to distinguish water-in-oil from oil-in-water emulsions. This behaviour is in the low frequency region of electromagnetic behaviour and represents the low frequency contribution to the dielectric spectrum. It is an aspect of electrochemistry. As the frequency is increased through to radio and visible radiation the way in which the energy is stored at the atomic and molecular level changes. However, the underlying relationship between the way in which the test signal is applied and the response is interpreted is unaltered. These interrelationships are important in, for example, the design of FTIR instrumentation but less commonly impinge on data analysis in the laboratory. This a distinct difference between mechanical and electromagnetic spectroscopy. Rheologists are interested in transforming the material responses from, say, the frequency domain to the time domain for very good reasons.

Firstly, it helps to provide a cross-check on whether the response of the material is linear or can be treated as such. Sometimes a material is so fragile that it is not possible to apply a low enough strain or stress to obtain a linear response. However, it is also possible to find non-linear responses with a stress/strain relationship that will allow satisfactory application of some of the basic features of linear viscoelasticity. Comparison between the transformed data and the experiment will indicate the validity of the application of linear models.

Secondly, rheological measurements can be applied to mimic the way materials are used and so relate to the chemical engineer's actual process. So whilst it may not be possible in practice to perform a measurement that mimics the process it may be possible to transform the data in such a way as to provide an insight into the way the material will behave in such a process. A typical example is encountered in the data handling of many rheometers which use the application of oscillating stress. The results are often reported as moduli rather than the measured property of compliance and this is achieved using a linear viscoelastic interrelationship.

Thirdly, in order to understand and predict the behaviour of our system we may wish to develop a chemical model for the behaviour. Certain experimental regimes are easier to model than others. By applying mathematical transformations to the models a wider range of experimental responses can be predicted.

Finally, and perhaps most importantly, is that many rheological processes operate over a very broad range of timescales, much broader than IR or UV spectral widths. The relaxation processes are often wider than those readily achievable by experimental measurement on any rheometer. We can extend our knowledge of the material to an extent beyond the operating ranges qualitatively by comparing data gathered in different tests with their transforms.

These interrelationships should not be viewed as an exercise in mathematics but an additional and useful tool in material characterisation. The mathematical relationships between some of the experiments are complicated to derive. The simplest example is presented in the next section and in sections that follow the relationships are stated rather than discussed in depth. They are summarised in Figure 4.18.

4.6.1 The Relationship Between Compliance and Modulus

When an oscillating stress is applied to a material the result is an

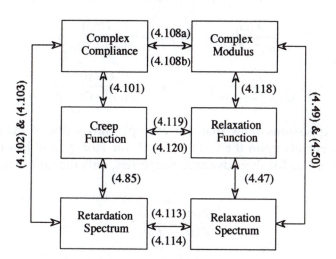

Figure 4.18 *The interrelationship between experiments and the spectra. The numbers in brackets refer to the equation numbers*

oscillating strain. The applied stress can be thought of as the independent variable and the strain response as the dependent variable. The ratio of dependent to independent variable defines the dynamic compliance. Similarly, when an oscillating strain is applied and a stress results the ratio of dependent to independent variable defines the complex modulus. Linear theory demands that these ratios are numerically the same. This can be represented algebraically as

$$\frac{\sigma^*}{\gamma^*} = \left(\frac{\gamma^*}{\sigma^*}\right)^{-1} \tag{4.105}$$

or in terms of modulus and compliance as

$$G^*(\omega) = \frac{1}{J^*(\omega)} \tag{4.106}$$

This relationship is the simplest of those describing the connection between stress and strain controlled measurements. Of course it also follows that complex terms must be similarly related:

$$G'(\omega) + iG''(\omega) = \frac{1}{J'(\omega) + iJ''(\omega)} \tag{4.107}$$

We can separate out the real and imaginary parts by multiplying by $(a - ib)$, as appropriate. This demonstrates the more sophisticated relationship between the complex compliances and the complex moduli:

$$G'(\omega) = \frac{J'(\omega)}{J'(\omega)^2 + J''(\omega)^2}; \quad G''(\omega) = \frac{J''(\omega)}{J'(\omega)^2 + J''(\omega)^2} \tag{4.108a}$$

$$J'(\omega) = \frac{G'(\omega)}{G'(\omega)^2 + G''(\omega)^2}; \quad J''(\omega) = \frac{G''(\omega)}{G'(\omega)^2 + G''(\omega)^2} \tag{4.108b}$$

These expressions show that, for example, any value of the storage modulus depends upon the values of both the storage and loss compliance. This is true for these dynamic properties, only the ratios interrelate simply:

$$\frac{G'(\omega)}{G''(\omega)} = \frac{J'(\omega)}{J''(\omega)} \tag{4.109}$$

Deriving this relationship is more than an exercise in mathematics and provides information on the nature of the material. We have argued that it is physically realistic for most viscoelastic systems to have a finite response to an instantaneous deformation or stress. There are some important consequences when considering the application of these ideas to the above expressions. As an example imagine that we have a viscoelastic solid material. We may relate the moduli and compliance as

$$G'_1(\omega) + G(0) + iG''_2(\omega) = \frac{1}{J_g + J'_1(\omega) + iJ''_2(\omega)} \tag{4.110}$$

where the subscripts 1 and 2 on the moduli refer to the storage and loss response excluding $G(0)$. Subscripts 1 and 2 on the compliance terms refer to the resulting response in compliance. Now as the frequency tends to zero the only modulus term to remain non-zero is $G(0)$. In Section 4.5.6 it was shown that the J'_1 in the limit of $\omega \to 0$ tends to J_e and J''_2 tends to zero. So as $\omega \to 0$ the result is

$$G(0) = \frac{1}{J_g + J_e} \tag{4.111}$$

This result is very interesting because whilst we have shown that $G(0)$ has been excluded from the relaxation spectrum H at all finite times (Section 4.4.5), it is intrinsically related to the retardation spectrum L through J_e. Thus the retardation spectrum is a convenient description of the temporal processes of a viscoelastic solid. Conversely it has little to say about the viscous processes in a viscoelastic liquid. In the high frequency limit where $\omega \to \infty$ the relationship becomes

$$G(\infty) = \frac{1}{J_g} \tag{4.112}$$

The high frequency elastic modulus does not appear in the retardation spectrum but is an intrinsic part of the relaxation spectrum. These features are reinforced when the interrelationship between the spectra are considered.

4.6.2 Retardation and Relaxation Spectrum

The expressions relating the storage and loss moduli to the relaxation spectra can be combined with those relating the complex modulus to the dynamic compliance. Although no proof has been provided here, the storage and loss moduli are linked by integral equations. Combining these relationships gives the following transforms:

$$L = \frac{H}{|A(\tau)|^2 + \pi^2 H^2} \tag{4.113}$$

$$H = \frac{L}{|B(\tau)|^2 + \pi^2 L^2} \tag{4.114}$$

with the integral expressions A and B given by

$$A(\tau) = G(0) - \int_{-\infty}^{+\infty} \frac{H(u)}{(\tau/u) - 1} \, d\ln u \tag{4.115}$$

$$B(\tau) = J_g - \frac{\tau}{\eta(0)} - \int_{-\infty}^{+\infty} \frac{L(u)}{1 - (u/\tau)} \, d\ln u \tag{4.116}$$

There are important points that can be learnt from these transformations. They reinforce what was stated in the previous section. For example $A(\tau)$ which relates H to L, contains $G(0)$ by addition to the integral in the relaxation spectrum, confirming the notion that at all finite times it does not contribute to H. In order to transform from L to H both J_g and viscosity are required. This hints at a general rule that can be applied to the form of the spectra and this will discussed in Section 4.7. The mathematical transforms themselves are quite awkward to determine numerically. You will notice that as the value of u approaches τ, so that their ratio approaches unity, the integrands diverge to positive or negative infinity! This 'singularity' requires care to negotiate when transforming experimental data.

4.6.3   The Relaxation Function and the Storage and Loss Modulus

These two experiments are fundamentally different in the nature of the applied deformation. In the case of the relaxation experiment a step strain is applied whereas the modulus is measured by an applied oscillating strain. Thus we are transforming between the time and frequency domains. In fact during the derivation of the storage and loss moduli these transforms have already been defined by Equation (4.53). In complex number form this becomes

$$G^*(i\omega) = G(0) + i\omega \int_0^\infty G(s) e^{-i\omega s} \, ds \tag{4.117}$$

For a viscoelastic liquid $G(0) = 0$. These expressions transform the stress relaxation function to the storage and loss moduli. Being Fourier transforms one can apply the Fourier inversion theorem to obtain the transform from the frequency to the time domain:

$$G(t) = \frac{2}{\pi} \int_{-\infty}^{+\infty} \left(\frac{G'(\omega)}{\omega} \right) \sin(\omega t) d\omega = \frac{2}{\pi} \int_{-\infty}^{+\infty} \left(\frac{G''(\omega)}{\omega} \right) \cos(\omega t) d\omega \quad (4.118)$$

4.6.4 Creep and Relaxation Interrelations

The relationship between creep and relaxation experiments is more complex. The complexity of the transforms tends to increase when stress and strain lead experiments are transformed in the time domain. This can be tackled in a number of ways. One mathematical form relating the two is known as the 'Volterra integral equation' which is notoriously difficult to evaluate. Another, and perhaps the conceptually simplest form of the mathematical transform, treats the problem as a functional. Put simply, a functional is a 'rule' which gives a set of functions when another set has been specified. The details are not important for this discussion, it is the result which is most useful:

$$p\mathbf{L}\{G(t)\} = (p\mathbf{L}\{J(t)\})^{-1} \quad (4.119)$$

Here **L** stands for the Laplace transform defined as

$$\mathbf{L}\{f(t)\} = \int_0^\infty f(t) \exp(-pt) dt \quad (4.120)$$

These two equations enable creep and relaxation to be related and complete our 'simple' combination of interrelations. The interrelations are summarised in Figure 4.18. The question we would really like to answer is how these interrelations apply to real systems. We can get an idea by looking at simple models. This is considered in the following section.

4.7 APPLICATIONS TO THE MODELS

The mathematics underlying transformation of the data from different experiments can be applied to simple models. In the case of the relationship between $G^*(\omega)$ and $G(t)$ it is straightforward. To give an example, consider a Maxwell model. It has an exponentially decaying modulus with time. We have indicated that the relationship between the complex modulus and the relaxation function is given by Equation (4.117). So if we substitute the relaxation function into this expression we get

$$G^*(i\omega) = G(0) + i\omega \int_0^\infty G(t)\exp(-i\omega t)dt = G(0)$$

$$+ i\omega \int_0^\infty G(\infty)\exp(-t/\tau_r)\exp(-i\omega t)dt \tag{4.121}$$

Now we can add the terms in the exponent together to obtain

$$G^*(i\omega) = G(0) + i\omega \int_0^\infty G(\infty)\exp[-t(1/\tau_r + i\omega)]dt \tag{4.122}$$

We should not be intimidated by the presence of $i = (-1)^{1/2}$, as the integral can be performed in a straightforward manner with i treated as a constant:

$$G^*(i\omega) = G(0) + i\omega G(\infty)\left[\frac{1}{-1/\tau_r - i\omega}\exp[-t(1/\tau_r + i\omega)]\right]_0^\infty \tag{4.123}$$

Putting in the limits and subtracting gives us

$$G^*(i\omega) = G(0) + \frac{i\omega G(\infty)}{1/\tau_r + i\omega} \tag{4.124}$$

This expression is the transform of the relaxation function. However, it is not in a readily recognisable form. If we multiply top and bottom of the quotient by τ_r we get

$$G^*(i\omega) = G(0) + \frac{i\omega \tau_r G(\infty)}{1 + i\omega \tau_r} \tag{4.125}$$

You will notice that this is the expression for a Maxwell model (see Equation 4.25). From Equations (4.121) to (4.125) we have applied a Fourier transform and confirmed that a Maxwell model fits at least this portion of the theory of linear viscoelasticity. The simple expression for the relationship between $J^*(\omega)$ and $G^*(\omega)$ allows an interesting comparison to be performed. Suppose we take our equations for a Maxwell model and apply Equation (4.108) to transform the response to an oscillating strain into the response for an oscillating stress. This requires careful use of simple algebra to give

$$J_1'(\omega) = \frac{1}{G} \tag{4.126}$$

$$J_2''(\omega) = \frac{1}{G\omega\tau} \tag{4.127}$$

If we were to use an oscillating stress in the equations for the mechanical analogues in Section 4.4.1 we would obtain these expressions. One should treat these transforms of the model responses with caution when applying the ideas to real materials. The most complete description of a material can be achieved by using the relaxation or retardation spectrum as our starting point. This allows access to all the linear viscoelastic experiments we can visualise. In order to do this we need to apply a novel function representing the spectrum. Imagine a distribution of relaxation processes represented by, for example, the log normal distribution in Equation (4.55). Now consider what happens as the half width at half height is reduced whilst we assume a constant $G(\infty)$ value. This means the area under the spectrum is constant (Equation 4.53). So as the peak gets narrower it gradually gets taller in order to maintain a constant area. In the limit one can visualise a 'peak' or a spike which is infinitely thin but which has an 'integratable' area of $G(\infty)$. Of course such a spectral function has great utility and a functional form can be defined which on integration has an area of unity. Such a function is termed the *Dirac delta function*, $\delta(x - x')$. The term in brackets $(x - x')$ is a method of representing the location of the function on the x axes. As defined it means that at all values of x it has a value of zero except at x'. So x' is the location of the spike on the x axis. This function is special and has some extremely useful properties which we will meet later in this text. Suppose we have a function $Q(t)$ where, for example, $Q(t)$ could be the relaxation function. An integral of the function multiplied by the Dirac delta function $\delta(t - t')$ gives the value at $Q(t')$:

$$\int Q(t)\delta(t - t')\mathrm{d}t = Q(t') \tag{4.128}$$

So suppose that we apply this property to our relaxation integral (Equation 4.47) such that the relaxation spectrum is replaced by a Dirac delta function at time τ_m:

$$H = G(\infty)\delta(\ln t - \ln \tau_m) \tag{4.129}$$

which when substituted in the integrand for relaxation gives us

$$G(t) = \int_{-\infty}^{+\infty} He^{-t/\tau}\mathrm{d}\ln\tau = \int_{-\infty}^{+\infty} G(\infty)\delta(\ln t - \ln \tau_m)e^{-t/\tau}\mathrm{d}\ln\tau = G(\infty)e^{-t/\tau_m} \tag{4.130}$$

This has now reduced the integral to a Maxwell relaxation process. In

other words a spike or delta function in the distribution of relaxation
processes represents a Maxwell model. It has a position in time which
defines the relaxation time. So it is not difficult to prove that two spikes
equate to two Maxwell models and so forth. However, this is not a
complete description. For a viscoelastic solid we know that we have a
$G(0)$ value. This can be represented by a Dirac delta function but only at
infinite time, which is an esoteric concept to say the least! Nonetheless the
fact that a Maxwell model is represented spectrally by a spike is an
important concept. A spike in the retardation spectra represents a Kelvin
model. It is possible to use these properties of the spectra and the Dirac
delta function to interrelate models. It is not straightforward however.
Throughout this chapter we have largely derived most of the mathematical functions used. Those that have not been derived are either
repetitious examples or require the application of complicated arith-
metic. For example a convolution integral is required to derive the
relationship between creep and relaxation and this has not been explored
in detail, as it is interesting but not essential to our argument. The
interrelationship between spectra has been well described by Gross.[2] For
a Kelvin model for example one can derive the characteristic relaxation
time and moduli of the springs in an equivalent Maxwell model and
accompanying spring. As the number of Maxwell or Kelvin models
increases, the calculations required to convert between relaxation and
retardation spectra increases. The mathematics is not particularly
involved but it is time-consuming. As the number of models increases
polynomials arise which inevitably can only easily be solved numerically.
There are useful rules that can be considered when modelling relaxation
and retardation spectra. These have been clearly stated by Gross. Put
simply these are:

- When one spectra (*i.e.* relaxation) is represented by a Dirac delta
  function, the other (*i.e.* retardation) is also a delta function.
- For a viscoelastic liquid the number of delta functions in the
  relaxation spectra always exceeds by 1 the number in the retardation
  spectra.
- For a viscoelastic solid the number of delta functions in the relaxation spectra is the same as the number in the retardation spectra.

4.8 MICROSTRUCTURAL INFLUENCES ON THE KERNEL

The previous sections contain a strong mathematical element, which is
inherent in the subject matter. Whilst it is possible to perform and model
viscoelastic behaviour without fully applying these ideas they can be

immensely powerful in predicting a range of experimental responses from simple models. A grasp of the area helps in understanding what one would expect from a material from a qualitative point of view. In the next chapter we will deal in detail with microstructural interpretation of linear viscoelastic theory, particularly using an exponential kernel. We illustrated near the beginning of this chapter that rheological behaviour could be related to the stress relaxation function (or its derivative, the memory function) and that this in turn was exponential in form. We borrowed mathematical terminology and described this exponential function as the kernel. There are some practical examples where an exponential function can be replaced by unusual kernels. There are two types of kernel that are used most frequently, the extended exponential and the power law kernel. The extended exponential has been applied to particulate systems with some ambiguity over its range of validity.[4] The power law kernel has been used to describe gelation behaviour.[5] In the following section we will consider these in detail.

4.8.1 The Extended Exponential

It is not that surprising that the linear viscoelastic response of many materials can be described using an exponential function, either as a sum or an integrated spectral response. The cynic would argue that ultimately we are undertaking multiple exponential fitting of the data. This enables us to fit a wide range of time decaying functions. The application of linear viscoelastic theory would inevitably allow other experiments to be described successfully. However, this does confirm that our original description of the process, *i.e.* the exponential kernel, is correct. One way of reconsidering this form of relaxation is to invoke a description of the relaxation mechanisms in terms of the molecular processes. This is considered in more detail in later sections, in which an exponential process can be seen to be appropriate for some systems. However, one can think about this problem phenomenologically. Consider a material such as a crystalline polymer composed of many randomly orientated grains. The atoms are ordered within the grains. If a strain is applied to this material the grains will deform slightly, changing the interatomic spacings, and when the strain is released the stress stored in the sample will relax. We can imagine that an individual grain relaxes as an exponential. We could reasonably suppose the rate at which the strain is relaxed depends upon the orientation of the grain since we would be deforming the grain along different crystal planes. Furthermore we could imagine that as the atomic orientations relax back to their quiescent configuration in each grain, local strains develop at the grain boundaries.

Put another way, as one grain relaxes the strain stored in its atomic bonds it deforms the neighbouring grain slightly. Of course every grain will influence every other, and it would be a highly co-operative process. The influence of one grain will be greatest on its neighbour and this will decay away with distance and time. A grain which in isolation would relax with a single Maxwell mode would relax somewhat more slowly as these modes couple together. This can be incorporated using a power law index n, usually less than unity:

$$G(t) = G(\infty)e^{-(t/\tau)^n} \tag{4.131}$$

This is an extended exponential. It operates within the remit of linear viscoelastic theory. So for example for a simple exponential we can show that the integral under the relaxation function gives the low shear viscosity:

$$\eta(0) = \int_{-\infty}^{+\infty} G(t)t\,\mathrm{d}\ln t = G(\infty)\tau \tag{4.132}$$

The same applies to the extended exponential. The integration is slightly more difficult but gives

$$\eta(0) = \int_{-\infty}^{+\infty} G(\infty)e^{-(t/\tau)^n}t\,\mathrm{d}\ln t = G(\infty)\tau\frac{\Gamma(1/n)}{n} \tag{4.133}$$

Here $\Gamma(n)$ is the mathematical relationship called the Gamma function, which can be calculated from tabulated data. The fact that the integral has a positive real solution indicates that the extended exponential represents a viscoelastic fluid unless an additional elastic process is included. Of course if we revert our thinking to our model for the process causing the coupling of the relaxation modes one can visualise other energy dissipation processes such as slip between grain boundaries. These will act to increase n towards unity. In practical terms it can be very difficult to distinguish between spectral processes and a slightly extended exponential.

4.8.2 Power Law or the Gel Equation[5]

In general it is fair to say that rheologists have been conservative in their use of non-exponential kernels. One particular form clearly stands out as a candidate for describing experimental data, at least for a limited range of relaxation times. This is the power law equation, often applied to gelling systems and consequently termed the *gel equation*. The relaxation function is described by a power law in time:

$$G(t) = St^{-m} \tag{4.134}$$

Therefore using the Boltzmann Superposition Principle (Equation 4.61) we have the gel equation:

$$\sigma(t) = S \int_{-\infty}^{t} (t - t')^{-m} \dot{\gamma}(t') dt' \tag{4.135}$$

This approach was pioneered by Winter[5] in terms of structural relaxation. The transforms already given in this chapter can be applied to this expression to give the following relationships:

$$G'(\omega) = \frac{\pi}{2\Gamma(m)\sin(m\pi/2)} S\omega^m \tag{4.136a}$$

$$G''(\omega) = \frac{\pi}{2\Gamma(m)\cos(m\pi/2)} S\omega^m \tag{4.136b}$$

Interestingly, the same power law governs both the relaxation and the dynamic moduli. Further use of this expression is given in Chapter 5 with a microstructural description. It is unrealistic to expect the gel equation to apply over the whole of the frequency and time ranges. For example it predicts an infinite high frequency modulus. However, it can apply over a wide range of frequencies and provide a useful description of the system.

4.8.3 Exact Inversions from the Relaxation or Retardation Spectrum

The Dirac delta function clearly provides one form of spectra which has an analytical transform to the viscoelastic experimental regimes discussed so far. An often overlooked function was developed by Tobolsky[6] and Smith.[7] They noted that particular forms of the relaxation or retardation spectra have exact analytical transforms. These functions give well defined spectra and provide good fits to experimental data. The relaxation spectrum is defined by the function:

$$H = G(\infty) \left(\frac{\tau}{\tau_0} \right)^{-m} e^{-\tau_0/\tau} \Gamma(m) \tag{4.137}$$

This has a corresponding relaxation function:

$$G(t) = G(\infty)\left(1 + \frac{t}{\tau_0}\right)^m \tag{4.138}$$

This function can be quite readily generalised as a summation of processes. This analysis remains a phenomenological approach. Bohlin[8] has suggested a generalisation of this approach indicating a relationship to a microscopic description of a system. Unfortunately this treatment has yet to be widely utilised. It is nonetheless tremendously useful for mapping data between spectra and experimental regimes.

4.9   NON-SHEARING FIELDS AND EXTENSION

So far we have concentrated on the application of shear fields to a sample. However, we can apply deformations by compressing or stretching a material. Indeed it is possible to visualise very complex combinations of strains and stresses. The response will relate to the bulk, shear and Young's relaxation functions in the appropriate form of the Boltzmann superposition integral. There are analogues to all the previously discussed responses. However, one note of caution should be heeded when relating bulk, shear and Young's moduli. This is that except in the low frequency limit the Poisson ratio is no longer single valued but is time dependent, so the relationship developed in Chapter 3 is not a correct description of the interrelationship for viscoelastic materials. For example the relationship between the Young's relaxation function $E(t)$ and $G(t)$ is given by[1,9]

$$E(t) = 2\left\{G(t) + \int_{-\infty}^{t} G(t - t')\dot{v}(t')\mathrm{d}t'\right\} \tag{4.139}$$

In order to proceed with the evaluation of the time-dependent Poisson ratio $v(t')$, both sets of relaxation behaviour are required. Now from Chapter 2 we know the Poisson ratio is the ratio of the contractile to the tensile strain and that for an incompressible fluid the Poisson ratio $v = 0.5$. Suppose we were able to apply a step deformation as we did for a shear stress relaxation experiment. The derivation then follows the same course as that to Equation (4.69):

$$E(t) = 2\{G(t) + vG(t)\} \tag{4.140}$$

This represents the relaxation of that stress in terms of the Young's relaxation function. Now we can express this as a viscosity by multiplying both sides by time t and integrating to get

$$\int_0^\infty E(t)t\mathrm{d}t = \int_0^\infty 2G(t)t\mathrm{d}t + 2\int_0^\infty vG(t)t\mathrm{d}t \qquad (4.141)$$

The first term on the left is the viscosity in extension, an extensional viscosity η_e:

$$\eta_e = 2\eta(0) + 2v\eta(0) \qquad (4.142)$$

So for an incompressible fluid the expression becomes

$$Tr = \frac{\eta_e}{\eta(0)} = 3 \qquad (4.143)$$

The constant Tr is called the Trouton ratio[10] and has a value of 3 in this experiment with an incompressible fluid in the linear viscoelastic limit. The elongational behaviour of fluids is probably the most significant of the non-shear parameters, because many complex fluids in practical applications are forced to extend and deform. Studying this parameter is an area of great interest for theoreticians and experimentalists.

4.10 REFERENCES

1. J.D. Ferry, *Viscoelastic Properties of Polymers*, 3rd edn, Wiley, New York, 1980.
2. B. Gross, *Mathematical Structures of the Theories of Viscoelasticity*, Hermann, Paris, 1968.
3. C.W. Mackosko, *Rheology, Principles, Measurements, & Applications*, Wiley-VCH, New York, 1994.
4. J.W. Goodwin and R.W. Hughes, *Adv. Coll. Interface Sci.* 1992, **42**, 303.
5. H.H. Winter, *Polym. Eng. Sci.* 1987, **27**, 1698; *Prog. Colloid Polym. Sci.* 1987, **75**, 104; *Macromolecules* 1988, **21**, 532.
6. A.V. Tobolsky, *J. Appl. Phys.* 1956, **27**, 673.
7. T.L. Smith, *J. Polym. Sci.* 1971, **C35**, 39.
8. L. Bohlin, *J. Coll. Interface Sci.* 1980, **74**, 423.
9. See reference 1 and N.W. Tschoegl, *The Phenomenological Theory of Linear Viscoelastic Behaviour*, Springer-Verlag, Berlin, 1989.
10. F. Trouton, *Proc. Roy. Soc. London* 1906, **A77**, 426.
11. M. Doi and S.F. Edwards, *The Theory of Polymer Dynamics*, Oxford University Press, Oxford, 1986.

CHAPTER 5

Linear Viscoelasticity II. Microstructural Approach

5.1 INTERMEDIATE DEBORAH NUMBERS $De \approx 1$

A rheological measurement is a useful tool for probing the microstructural properties of a sample. If we are able to perform experiments at low stresses or strains the spatial arrangement of the particles and molecules that make up the system are only slightly perturbed by the measurement. We can assume that the response is characteristic of the microstructure in quiescent conditions. Here our convective motion due to the applied deformation is less than that of Brownian diffusion. The ratio of these terms is the Péclet number and is much less than unity. In Equation (5.1) we have written the Péclet number in terms of stresses:

$$Pe = \frac{6\pi a^3 \sigma}{k_B T} < 1 \tag{5.1}$$

$k_B T/6\pi a^3$ is the 'thermal stress' and σ is the mechanical or rheological stress. To see the significance of Equation 5.1 we can consider three cases. If we have molecules such as cyclohexane we require $\sigma < 10^6$ Pa, whilst for a flexible polymer of moderate molecular weight a value of $\sigma < 100$ Pa would be satisfactory, but for a colloidal particle with a radius of 100 nm $\sigma < 0.2$ Pa would be required to ensure that the microstructure is relatively unperturbed.

In order to observe linear viscoelasticity, structural relaxation by diffusion must occur on a timescale comparable to our measurement time. The ratio of these times is the Deborah number. When this is of the order of unity our experiment will follow the relaxation processes in the material and the material will appear to be viscoelastic:

$$De = \frac{\text{structural relaxation time}}{\text{experimental measurement time}} \approx 1 \qquad (5.2)$$

Provided our experiment satisfies the conditions represented in Equations (5.1) and (5.2) the experiment is likely to be linearly viscoelastic. The advantages of such an experiment are two-fold. Firstly it allows us to compare and predict the response of the material to a wide range of linear viscoelastic experiments, thus allowing us to obtain information on a wide range of properties of the material. Secondly it is an indication of the state of matter in quiescent conditions as it can identify whether a material is solid- or liquid-like in nature and provide subtle distinctions in the phase behaviour. Rheology measures the state of matter and thus phase diagrams are important in establishing what form of rheology might be expected. The microstructural approach, which describes the rheological properties in terms of the spatial arrangement of particles and molecules, is complementary to the phenomenological approach. The linear viscoelastic relations can be used to transform the response to a strain to that for a stress for example. There are models for many systems so we will concentrate on the simple models that demonstrate the principles.

5.2 HARD SPHERES AND ATOMIC FLUIDS[1]

In order to understand how structure and particle forces affect rheology we might suppose that we should choose systems with the simplest form of interactions between the particles, and measure and model their behaviour. A good candidate for this class of materials would be the condensed inert gases, which have only weak interactions between the particles which can be treated as spherically symmetric. However, these systems are incredibly difficult to 'handle'. Moreover quantifying and tuning the forces between the particles is only readily achieved by either changing the gas or the pressure and temperature. Indeed the act of measurement alone can change the properties of the system. In order to support these experiments and understand more about the observed phenomena, computer simulations have been developed. In such 'computer experiments' any potential, density or temperature can be selected for our particle. Providing the 'physical rules' describing the response of the material which are used in the algorithm are correct we can subject this system to a range of changes. The simplest system we can visualise is a spherical particle that does not interact with its neighbour until it touches it. At this point the interaction energy rises to infinity and the particles can come no closer. The maximum concentration we can achieve with such a system is a volume fraction of approximately 0.74, equivalent to face centred cubic packing (fcc). We can perform a thought experiment where we place our particles in a box all moving with thermal motion and then progressively reduce the volume of this box. The volume fraction will rise and a corresponding increase in the pressure will be observed. We would expect ideal gas behaviour, so for n moles occupying a volume V the pressure is

$$P = \frac{n}{V}RT = \rho k_B T \qquad (5.3)$$

where ρ is the number of particles per unit volume. The viscosity is that of an ideal gas. As the size of the box is reduced further the finite volume occupied by the molecules needs to be accounted for and the pressure increases more rapidly than Equation (5.3) would predict. As we reduce the volume of the box still further a surprising feature is observed and this is shown in Figure 5.1.

Figure 5.1 shows that the pressure ceases to monotonically increase. The break point in the curve corresponds to a transition in the structural order. Instead of all the particles being distributed in a relatively disordered configuration, regions of the box begin to order in an fcc structure. The lower bound occurs when 49.4% of the volume is particles. This coexistence region extends along the plateau until at a critical volume the system is entirely composed of particles with an fcc order where the particles occupy 54.5% of the box volume. This is the

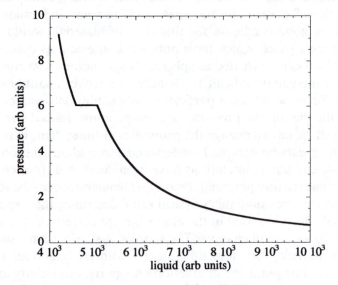

Figure 5.1 *The pressure versus volume curve for hard spheres*

order–disorder transition. The observation of this was as a result of the pioneering work of Alder and Wainwright.[2] The change is the result of the system finding that the fcc microstructural arrangement has a lower free energy than a disordered structure at the same volume. Effectively, the system avoids a further pressure increase by changing its microstructural arrangement. As the box is further reduced the fcc structure is compressed and the pressure again increases. The particles, whilst being constrained to a degree to fcc sites, still show on average an occupancy between these sites. Kincaid and Weis[3] showed that localisation to an fcc site did not occur until a particle volume occupancy of 63%. This limit is often equated to dense random packing. We might expect that the order–disorder transition would be accompanied by a change in rheology. In order to experimentally investigate the response of hard sphere systems we need to develop a near equivalent model system. Such a system may be described as quasi-hard sphere.

5.3 QUASI-HARD SPHERE DISPERSIONS

Many of the materials that we have a common everyday experience of, such as detergents and foodstuffs, show viscoelastic properties. Provided the stress or strain is low enough a linear viscoelasticity region can be observed where the stress changes in linear proportion to the strain. We can use the chemical properties of colloidal particles to 'tune' the pair potential energy of interaction between them. Quasi-hard sphere systems can be developed which display negligible van der Waals' attraction between a pair of particles and are very repulsive when they are nearly in contact. It might be supposed that these could be used to develop ideas and models for the rheological properties. Typically they are formed from non-aqueous dispersions although aqueous systems can be developed to show regions of near hard sphere behaviour. Two widely used particles are silica coated by an etherification reaction with octadecanol and poly(methyl methacrylate) coated by poly(12-hydroxystearic acid) as a hexamer. These are dispersed in solvents with a near refractive index match for the particles in order to reduce the attractive interactions. There is one major difference between these materials and true hard spheres and that is the presence of the dispersing medium. The medium will contribute to both the thermodynamic and hydrodynamic properties of the system. The first of these differences can be readily accommodated to an extent, but the second is still a major challenge. We will deal first with the thermodynamic issues.

5.3.1   Quasi-hard Sphere Phase Diagrams

In order to utilise our colloids as near hard spheres in terms of the thermodynamics we need to account for the presence of the medium and the species it contains. If the ions and molecules intervening between a pair of colloidal particles are small relative to the colloidal species we can treat the medium as a continuum. The role of the molecules and ions can be allowed for by the use of pair potentials between particles. These can be determined so as to include the role of the solution species as an energy of interaction with distance. The limit of the medium forms the boundary of the system and so determines its volume. We can consider the thermodynamic properties of the colloidal system as those in excess of the solvent. The pressure exerted by the colloidal species is now that in excess of the solvent, and is the osmotic pressure Π of the colloid. These ideas form the basis of pseudo one-component thermodynamics. This allows us to calculate an elastic rheological property. Let us consider some important thermodynamic quantities for the system. We may apply the first law of thermodynamics to the system. The work done in an osmotic pressure and volume experiment on the colloidal system is related to the excess heat adsorbed dQ and the internal energy change dE:

$$dQ = dE + \Pi \, dV \qquad (5.4)$$

For a reversible change the associated entropy change dS for the colloidal particles is given by applying the second law:

$$dQ = T \, dS \qquad (5.5)$$

We may combine these expressions:

$$dE = T \, dS - \Pi \, dV \qquad (5.6)$$

We can rewrite this in a new form at constant volume and entropy:

$$\left(\frac{\partial E}{\partial S} \right)_v = T \quad \text{and} \quad \left(\frac{\partial E}{\partial V} \right)_S = -\Pi \qquad (5.7)$$

The simplicity of the form of these expressions suggests that S and V are 'natural' variables of the internal energy $E(S,V)$. These expressions have demonstrated the first and second laws of thermodynamics for our colloidal system. However, it is not easy to perform an experiment at constant entropy on these systems. We can also express the Gibbs free energy $G(T,P)$:

$$G = E + \Pi V - TS \tag{5.8}$$

so that

$$\left(\frac{\partial G}{\partial T}\right)_{\Pi} = -S \quad \text{and} \quad \left(\frac{\partial G}{\partial \Pi}\right)_{T} = V \tag{5.9}$$

The Helmholtz free energy $A(T,V)$ in excess of that of the medium is

$$A = E - TS \tag{5.10}$$

so that

$$\left(\frac{\partial A}{\partial V}\right)_{T} = -\Pi \quad \text{and} \quad \left(\frac{\partial A}{\partial T}\right)_{V} = -S \tag{5.11}$$

The analogue to one-component thermodynamics applies to the nature of the variables. So A, S, U and V are all extensive variables, *i.e.* they depend on the size of the system. The intensive variables are Π and T – these are local properties independent of the mass of the material. The relationship between the osmotic pressure and the rate of change of Helmholtz free energy with volume is an important one. The volume of the system, while a useful quantity, is not the usual manner in which colloidal systems are handled. The concentration or volume fraction is usually used:

$$\varphi = \frac{V_{\mathrm{p}}}{V} \tag{5.12}$$

where V_{p} is the volume of all the particles in the system. This is also an extensive variable:

$$\frac{\mathrm{d}\varphi}{\mathrm{d}V} = -\frac{V_{\mathrm{p}}}{V^2} = -\frac{\varphi}{V} \tag{5.13}$$

This simple relationship allows us to express all the thermodynamic variables in terms of our colloid concentration. The Helmholtz free energy per unit volume depends upon concentration of the colloidal particles rather than the size of the system so these are useful thermo-dynamic properties. If we use a bar to symbolise the extensive properties per unit volume we obtain

$$\bar{A} = \bar{E} - T\bar{S} \tag{5.14}$$

and so the osmotic pressure becomes related to the concentration and the Helmholtz free energy per unit volume:

$$\Pi = \left(\frac{\partial \bar{A}}{\partial \varphi}\right)_T \tag{5.15}$$

What we would like to do is use these thermodynamic properties to calculate an equilibrium elastic moduli. The bulk modulus is by definition the constant of proportionality that links the infinitesimal pressure change resulting from a fractional change in volume (Section 2.2.1). In colloidal terms this becomes

$$K_T = \varphi \left(\frac{\partial \Pi}{\partial \varphi}\right)_T \tag{5.16}$$

This is the isothermal bulk modulus. Thus we can use our simulation data in Figure 5.1 and calculate a modulus for a hard sphere system. Equations (5.14) to (5.16) form an interesting hierarchy of equations:

$$K_T = \varphi \left(\frac{\partial \Pi}{\partial \varphi}\right)_T = \left(\frac{\partial^2 \bar{A}}{\partial \varphi^2}\right)_T = \left(\frac{\partial^2 \bar{E}}{\partial \varphi^2}\right)_T - \left(\frac{\partial^2 \{T\bar{S}^2\}}{\partial \varphi^2}\right)_T \tag{5.17}$$

The internal energy per unit volume relates to the Helmholtz energy, with its first derivative giving pressure and its second elasticity. We will see this form of relationship repeated again when we start to consider the specific nature of the particle arrangements. The osmotic pressure dependence as a function of volume fraction is shown in Figure 5.2. The isothermal bulk modulus is shown in Figure 5.3. The osmotic pressure shows an increase up to a volume fraction of 0.494 where the liquid-like order is replaced by two coexisting phases until a further increase where an ordered structure is formed.

The pressure curves are based on the calculations of Carnahan and Starling[4] (disordered phase) and Hall[5] (ordered phase). There is evidence that there is a long-lived metastable phase with liquid-like order determined by Woodcock, Pusey and van Megen[6] which extends the disordered branch to higher volume fractions diverging at 0.64. Experimental verification of these computer calculations has been obtained by Vrij *et al.*[7] They measured the osmotic compressibility of small silica particles in cyclohexane and found excellent agreement with the Carnahan–Starling model for compressibility up to $\varphi = 0.30$. In these models

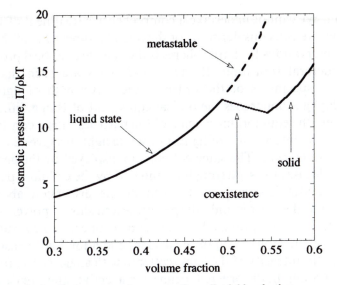

Figure 5.2 *The hard sphere osmotic pressure for colloidal hard spheres*

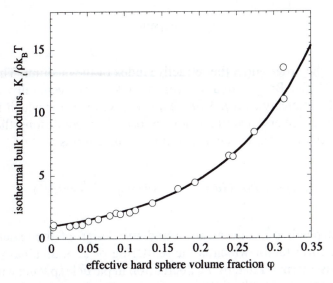

Figure 5.3 *The calculated hard sphere bulk modulus (line) compared with the data gathered by Vrij et al.[7] (shown as the points)*

the osmotic pressure and modulus did not include any hydrodynamic properties of the solvent and so these do not play a significant role in determining the bulk moduli. The osmotic compressibility was established by light scattering since both the compressibility and the scattered light are related to the structural arrangement of particles in the suspension. We cannot undertake a comprehensive treatment of scattering here but we can consider some of the main features. One of the most startling observations for hard spheres is the change in optical properties at the structural transition. If the particle sizes are such that their interparticle spacing is of the order of the wavelength of light then incident daylight undergoes the optical equivalent of Bragg diffraction. This is where the equilibrium spacing of the particles causes constructive and destructive interference of light. For white light this gives a range of colours or iridescence. The same colours are displayed by the gemstone opal; in both cases the scattering is dominated by the colloidal particles. Individual crystallites can be observed in the dispersion above the transition, and these grow and interpenetrate and show a polycrystalline nature at high concentrations. We can arrange an experiment such that we use just a single wavelength of incident radiation and measure the variation in the intensity of the light scattered with angle θ away from the transmitted beam. In the ordered regime constructive interference occurs at angles $\alpha = \theta/2$ for incident light of wavelength λ and layer spacing d_{hkl}:

$$m\frac{\lambda}{n} = 2d_{hkl}\sin(\alpha) \qquad (5.18)$$

where m is the order and n the refractive index of the medium. This is an analogue of the Bragg equation for X-rays. More generally we can consider the intensity as $I(q,\tau)$ where q is the wave vector which has the dimensions of inverse length. Under appropriate conditions the measured intensity is given by a combination of three terms:

$$I(q,\tau) = KP(q)F(q,\tau) \qquad \text{where } q = \frac{4\pi n}{\lambda}\sin(\theta/2) \qquad (5.19)$$

where K depends on concentration and the design of the experiment, $P(q)$ is the form factor which is the scattering of an individual particle and $F(q,\tau)$ is a term related to the fluctuation in particle spacing with time τ. This last term is called the intermediate scattering function. The intensity fluctuations are due to particle diffusive motion. At short times this is called the static structure factor $S(q)$, which is a measure of the equilibrium order in the suspension. It varies as a function of angle or wave vector, oscillating around unity. The peaks in the structure factor represent the primary location of the particles in reciprocal space. At zero angle, $i.e.$ $q = 0$, $S(q)$ relates directly to the isothermal bulk modulus of the dispersion:

$$K_T = \frac{\rho kT}{F(0,0)} = \frac{\rho kT}{S(0)} = \chi_T^{-1} \qquad (5.20)$$

where χ_T is the isothermal compressibility. This expression links the scattering, rheological and thermodynamic properties of the system. As yet we have not described either the viscous or the linear viscoelastic response of hard spheres in shear flows. This will be considered in the following section.

5.3.2 Quasi-hard Sphere Viscoelasticity and Viscosity

The correlation between experimental and theoretical values of the isothermal bulk modulus gives a good indication that the static properties of quasi-hard spheres are well understood. The viscous and viscoelastic behaviour when the system is above the dilute limit is far more difficult to predict. One reason for this is that the viscosity arises from complex hydrodynamic interactions. In a moderate to concentrated system these are multibody in nature. These result from the interactions between flow patterns on neighbouring particles and the extent of their coupling is theoretically difficult to determine. In addition the systems are 'near' hard spheres rather than 'true' hard spheres. There may be deviations in particle monodispersity, small amounts of residual charge at the surface, weak attractive forces present, the layer on the surface may modify the hydrodynamics away from a smooth surface or the particles may swell in the solvent. It is not practically possible to eliminate all these factors or to quantify them in enough detail. Moreover, as the concentration of the system increases, the particles are on average closer together and the deviations can become marked. Pragmatically we can overcome this difficulty by examining experimental data on all quasi-hard sphere systems and look for a commonality of behaviour. The Krieger–Dougherty[8] expression is a relationship that can describe the concentration dependence of the zero shear rate viscosity of quasi-hard spheres. It is discussed in detail in Section 3.5.2. It relates the viscosity to a packing fraction, at which point the structure becomes solid:

$$\eta(0) = \eta_0 \left(1 - \frac{\varphi}{\varphi_m(0)} \right)^{-[\eta]\varphi_m(0)} \qquad (5.21)$$

The intrinsic viscosity is the Einstein value $[\eta] = 2.5$ and the packing fraction $\varphi_m(0)$ is that in the low shear limit. As the volume fraction approaches the maximum packing fraction, the viscosity rapidly diverges. In order to utilise his expression we need $\varphi_m(0)$. One method is to fit the experimental data at low concentrations, say to the Einstein expression:

$$\eta(0) = \eta_0(1 + kc[\eta]) \qquad (5.22)$$

where c is the concentration of particles in units of mass per unit volume. The constant k is established by plotting a graph of the measured viscosity from capillary viscometry versus c. The slope divided by 2.5 gives k and the product kc gives ϕ_{hs}, the hard sphere volume fraction. The higher concentration curve is then fitted using Equation (5.21) in order to determine the maximum packing fraction of the particles. This *tends* to produce large values of packing, $\varphi_m(0) = 0.60$ and higher. Alternatively one can develop an effective hard sphere at high concentrations. Equation (5.21) can be rearranged to give

$$\frac{1}{[\eta]\varphi_m(0)} \ln\left\{\frac{\eta_0}{\eta(0)}\right\} = 1 - \frac{kc}{\varphi_m(0)} \qquad (5.23)$$

Plotting the left-hand side of the above equation versus c gives a y-axis intercept of unity and when $y = 0$ then $kc = \varphi_m(0)$. The values of k and $\varphi_m(0)$ can be adjusted iteratively to give the best linear fit. This tends to produce lower values of $\varphi_m(0) = 0.52$ close to the hard sphere transition. Structural relaxation in quasi-hard sphere systems has been investigated by diffusion measurements made using dynamic light scattering at a wide range of volume fractions. This has been performed by investigating the structure factor as a function of the wave vector q. The structure factor is related to the spatial arrangement of the particles. The peaks in $S(q)$ occur where the arrangement of particles in the dispersion cause a constructive interference of the light. We can estimate the position of the first peak in the structure factor, q_{max}, by equating the Bragg equation (Equation 5.18) with the wave vector equation (Equation 5.19):

$$q_{max} = \frac{2\pi}{d_{hkl}} \qquad (5.24)$$

This is a reasonable estimate for highly ordered systems. Each q is characteristic of a length scale within the dispersion. We can calculate $S(q)$ directly using an expression of Ashcroft and Lekener[9] based on a simple model for hard spheres by Percus and Yevick.[10] This model is shown in Figure 5.4 for a range of volume fractions. This is compared with some data of Ottewill[11] for $S(q)$ which was determined from neutron scattering for a typical quasi hard sphere system.

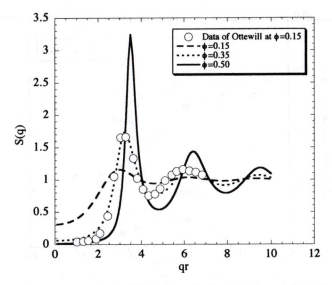

Figure 5.4 *The structure factor S(Q) determined using the method of Ashcroft and Lekener[9] compared with experimental data gathered by Ottewill[11] (points) using neutron scattering; r is the distance between particle centres*

The structure factor at q_{max} represents the nearest neighbour spacing of particles. So if the diffusive properties are measured at this peak we can follow the relaxation of the structure in the system through Brownian diffusion. We can examine the intensity on a range of time-scales. Most appropriate to structural relaxation is the timescale associated with the particles moving a distance which is a considerable fraction of the particle radius. This is the long time self-diffusion coefficient D_L. Pusey and coworkers[12] have found this to scale with the zero shear viscosity over a wide range of volume fractions. They found that

$$\frac{D_L(q_{max})}{D_0} = \frac{\eta_0}{\eta(0)} \tag{5.25}$$

where D_0 is the diffusion coefficient in the dilute limit. This suggests that the relaxation of the structure is determined by the low shear viscosity. This is not surprising because the particle is unlikely to undergo free diffusion and will experience the 'mean field' of all the surrounding particles. The corresponding relaxation time characteristic of the structure is given by the time to diffuse a distance equal to the particle radius:

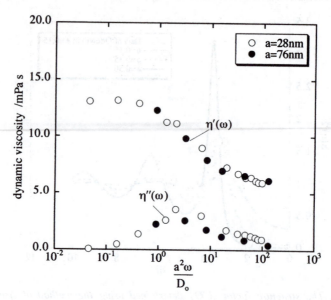

Figure 5.5 *The dynamic viscosity for a quasi-hard sphere dispersion from the data of Mellema et al.[13] The frequency has been normalised to the diffusion time for two different particle radii. The volume fraction is $\varphi = 0.46$*

$$\tau_r = \frac{a^2}{D_L(q_{max})} = \frac{6\pi\eta(0)a^3}{k_B T} \qquad (5.26)$$

There are not a great number of studies on the viscoelastic behaviour of quasi-hard spheres. The studies of Mellema and coworkers[13] shown in Figure 5.5 indicate the real and imaginary parts of the viscosity in a high-frequency oscillation experiment. Their data can be normalised to a characteristic time based on the diffusion coefficient given above.

These experiments suggest that as the long time self-diffusion coefficient approaches zero the relaxation time becomes infinite, suggesting an elastic structure. In an important study of the diffusion coefficients for a wide range of concentrations, Ottewill and Williams[14] showed that it does indeed reduce toward zero as the hard sphere transition is approached. This is shown in Figure 5.6, where the ratio of the long time diffusion coefficient to the diffusion coefficient in the dilute limit is plotted as a function of concentration.

This is strong evidence for assuming that dispersions of ideal hard spheres would be expected to show a transition in the viscous behaviour between $\varphi = 0.494$ and $\varphi = 0.545$. Also shown is the short time self-diffusion coefficient D_S. This still shows a significant value after the order–disorder transition. The problem faced by the rheologist in interpreting hard sphere systems is that at high concentrations there is

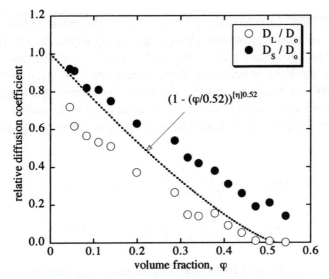

Figure 5.6 *The quasi-hard sphere diffusion data of Ottewill and Williams[14] as a function of volume fraction. The long (D_L) and short (D_S) time tracer diffusion coefficients are shown (symbols). The dotted line is 'representative' of the relative fluidity of a hard sphere dispersion*

an increased sensitivity to the deviations in ideality of the particles. In addition the multibody hydrodynamics still remain too difficult to evaluate completely. As the diffusion coefficient is inversely proportional to the viscosity it might be expected that the diffusive behaviour would be proportional to the relative fluidity. This is also plotted as the dotted line. An alternative approach is to deliberately build into the dispersion known interparticle forces that are significant relative to those due to the hydrodynamic effects. We shall consider this approach in the following section.

5.4 WEAKLY ATTRACTIVE SYSTEMS

The interparticle forces between some systems of colloidal particles can be adjusted such that any pair of particles in the dispersion are weakly attracted towards each other. The energy of the attraction must not be so great as to cause a permanent contact to form. Typically an energy of attraction of up $-20k_BT$ will allow the particles to remain in contact but be easily dispersed by gentle agitation. At very low levels of attraction, the interparticle energies cause the normal Brownian collisions to become modified. The particles dwell together as a pair longer and as a consequence the equilibrium structure in the dispersion becomes modified. This can in principle be detected by the scattering of radiation because the mean interparticle spacing between the particles will change. There will be a corresponding change in our structure factor $S(q)$. We would like to know how this change in both order and diffusive motion away from the hard sphere condition affects the relaxation behaviour, the elasticity, the pressure and the viscosity. To achieve this we can take our measured structure factor and calculate a quantity called the pair distribution function $g(r)$. The relationship between these quantities is given by:

$$g(r) = 1 + \frac{1}{2\pi^2 r \rho} \int_0^\infty [S(q) - 1]q\sin(qr)dq \qquad (5.27)$$

$$S(q) = 1 + \frac{4\pi\rho}{q} \int_0^\infty [g(r) - 1]r\sin(qr)dr \qquad (5.28)$$

In order to evaluate one from the other a numerical implementation is required. The pair distribution function is a tremendously useful quantity. If we consider a dilute suspension the particles are widely spaced with an average number density of particles given by ρ. As the concentration is increased and a material becomes more liquid-like, a short-range order develops, *i.e.* the particle concentration is no longer uniform but fluctuates about ρ. Suppose we freeze our system for a moment and select a test particle in our dispersion. If we examine the local concentration a radial distance r from the test particle we will find regions where the concentration $\rho(r)$ at r is less than the average and regions where it is greater than the average. As we travel away from the test particle, the local order reduces and $\rho(r)$ becomes closer to ρ. We can define our pair distribution or radial distribution function as

$$g(r) = \frac{\rho(r)}{\rho} \qquad (5.29)$$

and as $r \to \infty$ then $g(r) \to 1$. We do not always have the luxury of determining $g(r)$ or $S(q)$ experimentally directly on our system and so computer calculations are useful. These have been optimised against computer simulations by Henderson and the pair distribution function calculated in this manner is shown in Figure 5.7 for a range of volume fractions.[1]

The value of the peaks and troughs in the pair distribution function represent the fluctuation in number density. The peaks represent regions where the concentrations are in excess of the average value while the troughs represent a deficit. As the volume fraction is increased, the peaks and troughs grow, reflecting the increase in order with concentration. We

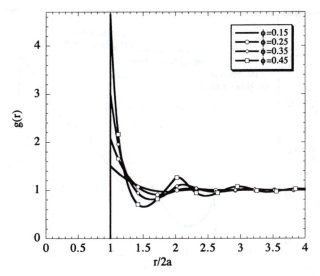

Figure 5.7 *The pair distribution function g(r) for hard spheres as a function of the dimensionless centre-to-centre separation. This was calculated using the algorithm from McQuarrie[1] at a range of volume fractions*

can measure this by calculating the number of particles around our test particle. If we multiply ρ by a volume we get the total number of particles contained in that volume. Since $g(r)$ is represented in terms of the radial distance from a central particle we need to calculate the number of particles in a spherical shell of thickness dr and area $4\pi r^2$, giving us the expression $g(r)4\pi r^2 dr$. This function is shown in Figure 5.8 and also oscillates through each peak and trough, reflecting a shell of coordinated neighbours.

The first nearest neighbour shell as a function of concentration is an important quantity, indicating how the structure develops locally. If z is the number of nearest neighbours we can sum up all the particles in the first nearest neighbour shell to the minimum position in $r^2 g(r)$ which marks the outer limit of that shell:

$$z = \int_{2a}^{\infty} 4\pi r^2 \rho g(r) dr = \frac{3\varphi}{a^3} \int_{2a}^{r_{min}} r^2 g(r) dr \tag{5.30}$$

In fact this basic idea allows us to intuitively connect the structure factor $S(q = 0)$ and the change in osmotic pressure. If we take $q = 0$ in Equation (5.28) we can see that it contains terms that are fluctuations in number concentration:

$$S(0) = 1 + \int_0^{\infty} 4\pi r^2 [\rho g(r) - \rho] dr \propto \frac{d\rho}{d\Pi} \tag{5.31}$$

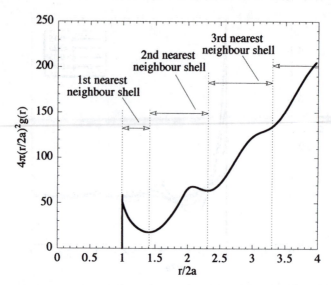

Figure 5.8 *A plot of the variation in the coordination shells $g(r)4\pi r^2$ with dimensionless separation for $\varphi = 0.45$*

These fluctuations lead to changes in the local compressibility of the structure, because this is proportional to the rate of change of pressure with concentration. As $S(0)$ results from the sum of these, the link to the isothermal bulk modulus can be understood in these terms. We would like to build into our nearest neighbour shell the effect of interparticle energy. So for example if we make our colloids attractive this will on average increase the number of nearest neighbours relative to the hard sphere value. From inspection of Equation (5.30), it is clear that this can only be reflected by modifying the pair distribution function. This will affect the structure factor, so ideally *all* we need to do is measure $S(q)$, calculate the structure represented by $g(r)$ and determine the rheology. This would require a great deal of experimental characterisation, but we can circumvent this if we can modify our computer calculations for the hard sphere pair distribution function. There are various schemes for doing this. Consider the pair potential shown in Figure 5.9. This is for a 500 nm radius polystyrene latex particle coated with a surfactant. The particle has a ζ potential of -9.7 mV in 0.5M sodium chloride.

There is a weak secondary minimum in the energy curve of about $-7k_BT$ that arises from the attractive force between the particles. Past this minimum the energy rises again as a result of the repulsive force between particles of similar charge. As two particles approach they will nearly touch and tend to dwell in the minimum of the energy distance curve. The average energy for two particles moving apart along a common axis with Brownian motion is k_BT but collisions cause fluctua-

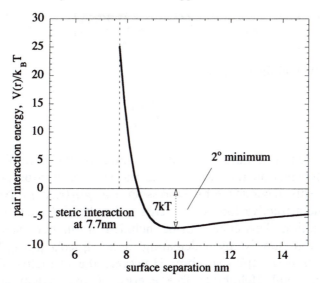

Figure 5.9 *The pair interaction potential between two polystyrene particles (radius 500 nm) in 0.5 M electrolyte. These were coated by a short chain surfactant of a length of 3.8 nm*

tions around this value. Thus some particles will receive enough energy to escape from this minimum and to either separate or to approach closer than the secondary minimum distance. Those that approach closer than a distance σ will be subject to a strong repulsive force. We can estimate the probability that particles approach closer than the minimum position using a Boltzmann energy distribution:

$$P[V(r)] = \frac{\exp(-V(r)/k_B T)}{Z} \tag{5.32}$$

The term Z here is the partition function and represents a sum over all the states. It ensures that a sum of all the probabilities is unity. In order to achieve a close separation between the particles they must possess an increasing energy. Clearly from Equation (5.32) an exponentially reducing number will achieve this. We can imagine there exists a separation which so few particles ever achieve that this effectively represents the closest distance a pair of particles can approach. This gives an effective hard sphere diameter. There are various perturbation schemes for defining where this might occur. One such is the Barker–Henderson perturbation (BH). The basic notion is that a small perturbation from the hard sphere potential can be incorporated by a simple modification of existing hard sphere models. The effective diameter or Barker–Henderson diameter is given by

$$d_{BH} = 2a + \int_{2a}^{\sigma} \left[1 - \exp\left(\frac{-V(r)}{k_B T}\right)\right] dr \qquad (5.33)$$

giving a new volume fraction

$$\varphi_{BH} = \varphi \left(\frac{d_{BH}}{2a}\right)^3 \qquad (5.34)$$

We have seen in Figure 5.7 that the value of the pair distribution function depends on the radius a and the volume fraction φ. We can emphasise this dependence by expressing it as $g_{HS}(r, 2a, \varphi)$. Now as we increase the attractive forces we increase the effective particle size and volume fraction. This effect can be included in our new pair distribution function simply by calculating it as $g_{HS}(r, d_{BH}, \varphi_{BH})$ and it represents a new repulsive core. However, the attractive force will modify the total Helmholtz free energy of the system and, from Equation (5.17), the compressibility and internal energy. Provided it is relatively small we can add in the perturbing energy, subsequent changes in order and local compressibility. This allows for the attractive energy. The new pair distribution function for particles with a 'softer' interaction becomes

$$g(r, 2a, \varphi) = g_{HS}(r, d_{BH}, \varphi_{BH})[1 - \Phi(r)\rho k_B T \chi_T] \qquad (5.35)$$

The term $\Phi(r)$ is a pair potential that contains only the attractive potential, because the repulsion effects have been allowed for by the effective volume fraction and hard sphere diameter. The new potential can be defined as

$$\frac{\Phi(r)}{k_B T} = \begin{cases} \infty & r < d_{BH} \\ V_{min}/k_B T & d_{BH} \leq r \leq r_{min} \\ V(r)/k_B T & r_{min} < r \end{cases} \qquad (5.36)$$

We have introduced a statistical mechanical approach, illustrating how the material properties and rheology play a role at the microscopic level. Our main reason for doing this is to determine the microstructure and calculate the macroscopic rheological properties. We can now evaluate the coordination number z from Equation (5.30) for our colloid pair potential in Figure 5.9. The variation of z with volume fraction is shown in Figure 5.10.

The first thing to note is the sigmoidal form of the curve; as the concentration is increased the coordination number increases. As the

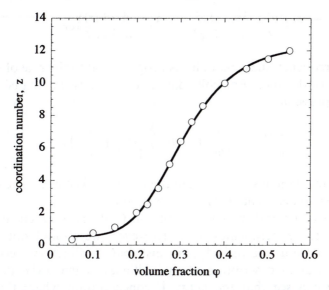

Figure 5.10 *The nearest neighbour coordination number calculated for the system with the pair potential in Figure 5.9*

volume fraction reaches $\varphi = 0.5$ the coordination number approaches 12. This is the maximum number of nearest neighbours that can be achieved in a close packed structure. Here we are approaching the hard sphere transition from liquid-like to solid-like order. At around $\varphi = 0.25$ the number of neighbours approaches 3, and this is the minimum number required to build a network of particles. At this concentration we can visualise the particles in the dispersion building a connected structure. This concentration coincides with a marked change in the elastic properties. We can determine these from the microstructure in the dispersion. If we multiply the pair interaction energy per particle $V(r)/2k_{\mathrm{B}}T$ by the number of particles at any separation $4\pi r^2 g(r)\mathrm{d}r$ we can define the internal energy per unit volume or internal energy density in dimensionless form:

$$\bar{E} = \frac{\bar{E}a^3}{k_{\mathrm{B}}T} = \frac{9\varphi}{8\pi} + \frac{3}{2}\varphi \int_0^\infty r^2 g(r) \frac{V(r)}{k_{\mathrm{B}}T}\mathrm{d}r \qquad (5.37)$$

This forms one of a hierarchy of equations we can write, representing the properties of the dispersion in terms of the microstructure. We can apply this idea to the osmotic pressure by considering the force acting on a pair of particles. The first derivative of energy with respect to distance provides us with force:

$$\frac{\Pi a^3}{k_B T} = \frac{3\varphi}{4\pi} - \frac{3\varphi^2}{8\pi a^3} \int_0^\infty r^3 g(r) \frac{\mathrm{d}}{\mathrm{d}r}\left(\frac{V(r)}{k_B T}\right) \mathrm{d}r \qquad (5.38)$$

The high frequency shear modulus is proportional to the rate of change of force with distance. For colloidal systems this is dominated by the integral expression[15]

$$\frac{G(\infty)a^3}{k_B T} = \frac{3\varphi^2}{40\pi a^3} \int_0^\infty g(r) \frac{\mathrm{d}}{\mathrm{d}r}\left[r^4 \frac{\mathrm{d}}{\mathrm{d}r}\left(\frac{V(r)}{k_B T}\right)\right] \mathrm{d}r \qquad (5.39)$$

Thus for our colloidal system we can calculate the elasticity and compare this with experimental data. This is shown in Figure 5.11.

The microstructural model in Equation (5.39) is in excellent agreement with the experimental data. However, we should not readily conclude that our perturbed pair distribution function necessarily represents a good description of the colloidal system over all length scales. We can see that the network concentration where the shear modulus becomes significant is around $\varphi = 0.25$, another good indication that the model represents the onset of networking in the system. However, the second derivative of energy with respect to distance dominates Equation (5.39). This reduces very rapidly with distance. Once we are at a distance of r that is greater than the nearest neighbour shell the integral effectively reduces to zero. In other words the elasticity

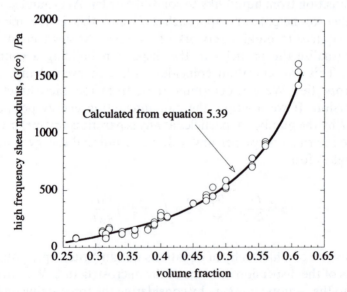

Figure 5.11   *The shear modulus obtained from experiments compared with the calculated value from Equation (5.39). The system is the same as that used to calculate the pair potential in Figure 5.9*

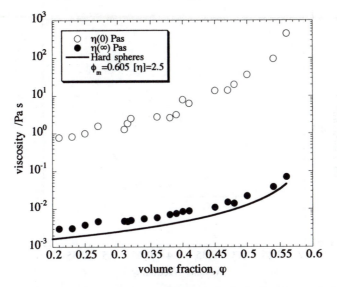

Figure 5.12 *The measured viscosity as a function of volume fraction at the low and high shear rate limits. This is the data for the system with the pair potential shown in Figure 5.9*

is dominated by nearest neighbour interactions. Thus whilst our pair distribution function may provide a good approximation to the structure at short length scales we do not know how well it corresponds to the system over longer length scales. One way of considering this is to measure the viscosity and compare this to theoretical calculations for viscosity in the low stress limit. The data shown in Figure 5.12 for the viscosity show a very rapid increase with concentration. The longest relaxation time, given by the ratio of the low shear stress viscosity to the high frequency elastic moduli given in Figure 5.13, indicates a more rapid increase in relaxation time once the hard sphere transition is approached. A similar feature is observed in the extrapolated Bingham yield stress (Section 6.3.4).

The major difficulty in predicting the viscosity of these systems is due to the interplay between hydrodynamics, the colloid pair interaction energy and the particle microstructure. Whilst predictions for atomic fluids exist for the contribution of the microstructural properties of the system to the rheology, they obviously will not take account of the role of the solvent medium in colloidal systems. Many of these models depend upon the notion that the applied shear field distorts the local microstructure. The mathematical consequence of this is that they rely on the rate of change of the pair distribution function with distance over longer length scales than is the case for the shear modulus. Thus

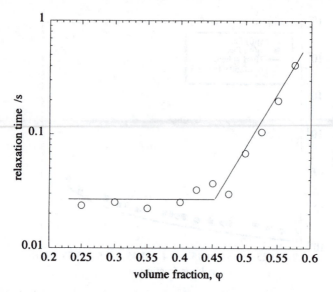

Figure 5.13 *The relaxation time versus volume fraction for a system with the pair potential shown in Figure 5.9*

we need to generate pair distribution functions that are precise over long length scales and are representative of the details of the curvature of $g(r)$. Possibly only computer simulations will enable this to be achieved. There are also limitations to the applicability of a perturbation scheme as a method of including the colloid pair potential. There are simple indications of this, for example the maximum packing that can be achieved with spheres is 74%. When the attraction is high between the particles the perturbation scheme can readily predict values that exceed this. Under such circumstances where a reliable prediction of the structure is not possible the shear modulus can be represented in a power law form:

$$G(\infty) = A\varphi^m \qquad (5.40)$$

where $m = 3$–4. This range of values is appropriate to the data in Figure 5.11.

5.5   CHARGE REPULSION SYSTEMS

Charged colloidal spheres can be produced using both organic and inorganic particles. In low electrolyte conditions, the interparticle forces are dominated by the charge repulsion forces between the particles. Interrogation of the microstructure using scattering techniques indicates that a phase change occurs at a critical concentration. This displays a structural change akin to the order–disorder transition seen with hard spheres. This occurs at lower volume fractions than with hard spheres. At low particle volume fractions and low ionic strengths the initial transition tends to a body centre cubic order (bcc). Further increases in concentration rapidly take the structure toward fcc order. There can be zones of coexistence between ordered and disordered structures as seen with hard spheres. The transition is accompanied by a change in the rheology of the systems, as was shown by Lindsay and Chaikin.[16] The transition between liquid-like and solid-like behaviour has been correlated to the dimensionless energy density. As we have seen with weakly attractive systems we can describe the internal energy, osmotic pressure and high frequency elastic modulus by a hierarchy of integral equations. These expressions still hold for charge-repulsive interactions. However, once the ordered state has been achieved we can visualise the colloidal particles constrained to a lattice site. The nearest neighbours dominate the interactions between the particles and are the most significant term in controlling the shear modulus. The centre-to-centre separation for a given order and packing R is given by

$$R = 2a \left(\frac{\varphi_m}{\varphi} \right)^{1/3} \tag{5.41}$$

where φ_m is the maximum packing, which for fcc is 74% and for bcc is 68%. Once the system achieves an ordered structure the number of nearest neighbours z remains invariant with an increase in concentration. The coordination number for fcc is $z = 12$ and for bcc is $z = 10$. This is an important distinction between systems dominated by charge repulsion and those which are weakly flocculated. Above the transition in the former case the coordination number is constant with volume fraction but the interparticle spacing reduces. In contrast, with weakly flocculated systems the interparticle spacing is set by the position of the minimum of the energy well and is invariant but the coordination number increases with increasing concentration. We can replace the term determining the number of neighbours $4\pi r^2 g(r)$ by the Dirac delta function at R (see Section 4.7 for a discussion of its properties):

$$\rho 4\pi r^2 g(r) \cong \delta(r - R)z \tag{5.42}$$

This allows the integrals representing the relationship between structure and macroscopic properties to be simplified to their value at R. This represents a cell model and so for the appropriate property:

$$\bar{E} = \frac{\bar{E}a^3}{k_B T} = \frac{9\varphi}{8\pi} + \frac{3\varphi z}{8\pi} \frac{V(R)}{k_B T} \tag{5.43}$$

$$\frac{\Pi a^3}{k_B T} = \frac{3\varphi}{4\pi} - \frac{\varphi z R}{8\pi} \frac{d[V(R)/k_B T]}{dR} \tag{5.44}$$

$$\frac{G(\infty)a^3}{k_B T} = \frac{\varphi z R^2}{40\pi} \frac{d^2[V(R)/k_B T]}{dR^2} \tag{5.45}$$

We can also develop a zero Kelvin lattice model with slightly different spatial averaging. This gives[17]

$$\frac{G(\infty)a^3}{k_B T} = \frac{3\varphi z R^2}{256} \frac{d^2(V(R)/k_B T)}{dR^2} \tag{5.46}$$

Now if we examine the available experimental data for charged systems the rheological properties show a transition with increasing volume fraction. For a given system we observe that the viscosity diverges and becomes too great to be measured past a critical concentration (Chapter 3). If the energy density where this occurs is plotted against volume fraction, a 'phase map' can be developed where there is a transition between measurable viscosity at low shear rates and apparently solid-like behaviour. This is shown in Figure 5.14, which was constructed from studies on 11 different charged systems.[18]

Also shown is the dimensionless energy density for hard spheres, $3k_B T/2$ per particle. The viscoelastic liquid zone is difficult to define

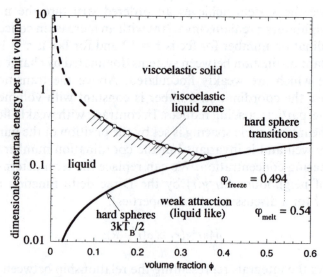

Figure 5.14 *The dimensionless internal energy versus volume fraction, indicating empiri-cally defined zones of liquid-like and solid-like behaviour*

clearly but occurs as a narrow boundary close to the region of solid-like behaviour. The solid line for the 'liquid–solid' transition is defined empirically by

$$\bar{E} = 0.05/\varphi \qquad \text{for } 0.1 < \varphi \leq \left(\frac{2\pi}{45}\right)^{1/2} \qquad (5.47)$$

The lower volume fraction limit is set by the range of experimental data, and the upper limit is set by the hard sphere internal energy $3k_BT/2$. In principle, a prediction of the order–disorder transition should be possible by calculating the free energy of the two states as a function of concentration. The system with the lowest free energy would be the favoured one. Russel *et al.*[19] suggested a simple approximation based on an effective diameter for the particles. He argued that the order–disorder transition seen in charged repulsive colloids was analogous to the hard sphere transition. The charged colloids would have an effectively larger diameter and hence volume fraction. The transition would occur when this effective volume fraction achieved $\varphi = 0.5$. In order to establish the effective hard sphere diameter Russel examined the integrand in the Barker–Henderson expression (Equation 5.33) and noted that it was a very rapidly changing function of distance. So for the typical form of repulsive potential at low electrolyte concentrations and particle radii we can substitute into the expression the repulsive pair potential:

$$\frac{V(r)}{k_BT} = \alpha \frac{\exp(-\kappa r)}{\kappa r} \qquad (5.48)$$

where κ is the Debye–Hückel length proportional to the electrolyte concentration of the dispersion, and α is related to the properties of the particle and the solvent (Section 3.5.4). The integrand changes very rapidly from zero away from the particle surface to unity at some distance r_0. This allows r_0 to be used as an effective hard sphere diameter. We can write this as

$$1 - \exp\left(\frac{V(r)}{k_BT}\right) = \left\{ \begin{matrix} 1 & r \leq r_0 \\ 0 & r > r_0 \end{matrix} \right\}$$

The separation at which this occurs can be taken to be when the pair potential and Brownian energy equate, *i.e.* $V(r_0) = 1k_BT$. This gives

$$r_0 \frac{1}{\kappa} \ln\{\alpha/\ln[\alpha/\ln(\alpha/\ldots)]\} \qquad (5.49)$$

This diameter can be used to estimate the viscosity (see Section 3.5.4). Our interest is in determining where the hard sphere boundaries occur:

$$0.50 > \varphi\left(\frac{r_0}{2a}\right)^3 \qquad \text{liquid-like order} \qquad (5.50a)$$

$$0.55 < \varphi\left(\frac{r_0}{2a}\right)^3 \qquad \text{solid-like order} \qquad (5.50b)$$

We can compare this to the expression derived from experimental data for transition in internal energy density. Good agreement between the empirical model and the experimentally derived curve is observed (Figure 5.15).

Figure 5.15 is a plot of the the value of volume fraction against κa. The sigmoidal shape reflects the increase in volume fraction required to achieve the *same* energy of interaction between the particles with increasing κa. It displays a plateau because at high ionic strengths the charges on the particles become fully shielded from each other and the particles approach the hard sphere order–disorder transition value. Russel showed that this also acts as a good prediction of the optical changes seen as particles achieve an ordered structure. In Figure 5.15 the transition observed by Hachisu, Kobayashi and Kose[20] is also compared

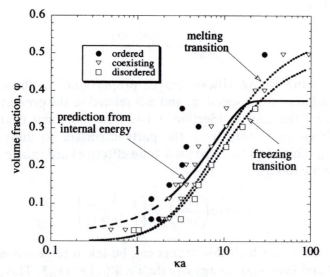

Figure 5.15   *A comparison with the model due to Russel et al.[19] and the internal energy at the viscoelastic solid transition. Also shown from data points is the order–disorder transition observed by Hachisu et al.[20]*

with the two approaches and confirms the notion that the onset of linear viscoelasticity accompanies the phase change.

An equilibrium model may not be representative of the true situation commonly faced in the laboratory. The relaxation behaviour of the samples becomes progressively longer with increasing volume fraction. It is quite reasonable to suppose that, at high particle concentrations and low electrolyte concentrations, the relaxation times become so long that it is impractical to allow all the stresses and strains to relax from the sample prior to measurement. Stress relaxation studies for a range of particles that show nearly complete relaxation is shown in Figure 5.16.[21]

The data has been superimposed by dividing the relaxation function $G(t)$ by $G(t = 0)$, the limiting short time value, and the time has been divided by the characteristic relaxation time τ_r. The first feature to notice is that the stress relaxation function overshoots and shows a peak. This is an example of non-linear behaviour. It is related to both the material and the instrumental response (Section 4.5.1). The general shape of the curves (excluding the stress overshoot) can be described using two approaches.

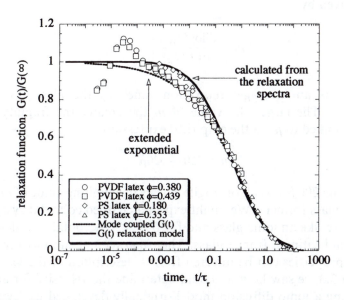

Figure 5.16 *Typical stress relaxation data for concentrated charge dispersions. Two models are shown, one based on a model for the relaxation spectra (Equation 5.59) and one based on an extended exponential (Equation 5.51)*

The first we will consider is based on the Kohlrausch law, which for stress relaxation is given by

$$G(t) = G(\infty) \exp\left[-(t/\tau)^{\beta}\right] \tag{5.51}$$

This is an extended exponential and is discussed further in Section 4.8.1. For the data in Figure 5.16 the value of β is 0.35. This expression implies the material is a viscoelastic liquid since $G(t)$ reduces to zero at long times. The extended exponential has been used to describe the behaviour of glass forming materials. Glasses can be quantified as being strong or fragile. Fragile glass formers show strong deviations away from Arrhenius behaviour (Chapter 2) and possess shorter range order. Molten salts are typical examples of fragile glasses where there is no directionality to the 'ion bonding'. Stronger glass formers are Arrhenius in nature and are typified by the presence of directional bonding. The viscosity or relaxation time as a function of temperature can be used to establish this classification for molecular and ionic species. However, for charged colloidal species in an aqueous environment there is unlikely to be a simple relationship to the temperature simply because so many of the system parameters depend upon it. Bohmer *et al.*[22] have assembled values for β for a wide range of systems. Their data suggests a relationship between β and the fragility. The fragility is represented by a term m which is given by

$$m = \left.\frac{\mathrm{d}[\log(\tau_{\mathrm{ave}})]}{\mathrm{d}(T/T_{\mathrm{g}})}\right|_{T=T_{\mathrm{g}}} \tag{5.52}$$

where τ_{ave} is an 'average' relaxation time, T_{g} the glass transition temperature. The larger the value of m the greater the fragility. The fragility is linked to β via the empirical expression:

$$m = 250 - 320\beta \tag{5.53}$$

m varies from 200 for weak or fragile glasses to a limiting value of $m = 16$ for strong glass formers. We might expect a charged colloidal system to behave more like an ionic glass and for $\beta = 0.35$ we obtain a calculated $m = 140$, indicating a fragile glass. The experimental support for some glass-like properties can be inferred from a description of the viscosity. In Section 3.3 we saw how we could determine the viscosity for atomic liquids using a jump diffusion model originally developed by Eyring.[23] The same idea can be extended to colloidal systems at a concentration below the viscoelastic liquid–solid transition.[24] The result is

$$\eta(0) = \eta_0 \left(1 - \frac{\varphi}{\varphi_m}\right)^{-[\eta]\varphi_m} + \frac{h}{R^3}\exp(E^*/k_B T) \tag{5.54}$$

where h is Planck's constant. The first term on the right represents the hydrodynamic contribution to the flow, the second term the pair potential contribution. This model assumes that the dispersion consists of particles in a well ordered arrangement with an average spacing given by fcc packing. However, the lattice contains defects in the form of vacancies. Flow is achieved by particles 'jumping' from one site to the next. The activation energy E^* is that associated with a particle making the jump. The action of the 'jump' diffusion results in one particle passing an energy barrier created by its neighbours. Figure 5.17 shows a plot of the viscosity versus concentration for the modified Eyring model given in Equation (5.54) compared with experimental data.

This curve highlights the transition between the liquid-like and solid-like response. The form of the curve is such that there is a very rapid transition between the zones, with the viscosity changing by nearly six orders of magnitude for a small change in volume faction. This highlights the difficulty of the experimental investigation of the transition between liquid-like and solid-like responses. At high enough concentrations the material is effectively solid-like on the timescales of colloid relaxation. The elastic modulus changes less rapidly and this is shown in Figure 5.18.

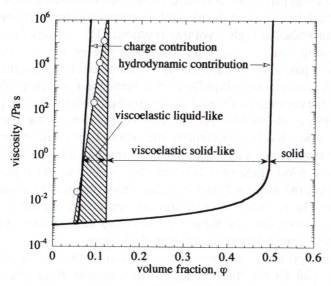

Figure 5.17 *The low shear viscosity as a function of concentration for a latex particle with an electrolyte concentration of $10^{-4}M$ and radius of 38 nm. This is compared with the modified Eyring[23] model and the hydrodynamic contribution to the flow*

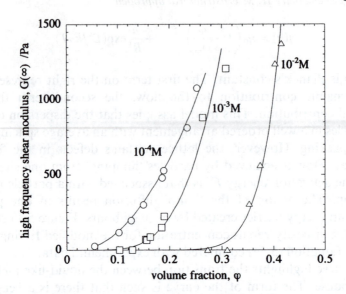

Figure 5.18 *The high frequency shear modulus versus volume fraction for a polystyrene latex for three different electrolyte concentrations. The symbols are the experimental data and the solid lines are calculated fits using a cell model. The radius of the latex particles was 38 nm*

The shear modulus has been calculated from the lattice model (Equation 5.46) for a polystyrene latex dialysed against three different electrolyte concentrations. It is clear from both the calculations and the experimental data that, as the electrolyte concentration is increased, the elasticity increases at higher volume fractions. This reflects the effect of the increasing ionic strength as it reduces the energy of interaction between the particles. In addition at higher electrolyte concentrations, the moduli increase more rapidly with volume fraction. This reflects the change in the curvature of the force–distance curve between the particles with increasing electrolyte. As the volume fraction of the particles is increased at low levels of electrolyte the effect of the counter ions which dissociate from the surface of the particles must be allowed for in order to achieve a good agreement between the model and the experimental data. The charge-stabilised particles that show the most gradual change of shear modulus with concentration are those that are most likely to display a measurable low shear rate viscosity and a stress relaxation profile which completely relaxes the applied strain.

Another approach we can use to describe the stress relaxation behaviour and all the linear viscoelastic responses is to calculate the relaxation spectrum H. Ideally we would like to model or measure the microstructure in the dispersion and include the role of Brownian diffusion in the loss of structural order. The intermediate scattering function contains information appropriate to this calculation. However, the long relaxation times can make it impractical to measure these properties on appropriate timescales. We can make a simple approximation for the relaxation spectra however by assuming the dispersion is polycrystalline, formed from many randomly orientated grains each with an fcc structure. The measured shear modulus $G(\infty)$ is the value for an 'average' grain. At any instant in time some grains will have a higher local particle concentration and hence a higher shear modulus, and of course some will have a lower local particle concentration and so have a lower modulus. The energy opposing close approach of the particles is such that only a few will be more elastic so to a first approximation we can suppose that relaxation occurs by an individual grain losing its elasticity by particles diffusing from R to a distance $R(\tau)$ at time τ. The relaxation spectrum can be represented by the product of the fluctuation in the elastic moduli of a configuration given by $\Delta G(\tau)$ multiplied by the probability of the configuration occurring $P[\Delta U(\tau)]$:

$$H(\tau) = \Delta G(\tau) P[\Delta U(\tau)] \tag{5.55}$$

The probability of a fluctuation occurring is given by a Boltzmann distribution in the pair potential fluctuation. The energy associated with a fluctuation $\Delta U(\tau)$ is given by the fluctuation in the separation:

$$\Delta U(\tau) = V[R(\tau)] - V(R) \tag{5.56}$$

The fluctuation in elasticity is given by the difference between the elasticity in the high frequency limit and the elasticity of the new configuration:

$$\Delta G(\tau) = G(\tau) - G(\infty) \tag{5.57}$$

Here $G(\tau)$ is the modulus that results from the particle diffusing a distance $R(\tau)$. This can be represented by a cell model with a time dependent separation:

$$G(\tau) = \frac{\alpha}{R(\tau)} \frac{\mathrm{d}^2 V[R(\tau)]}{\mathrm{d}R(\tau)^2} \tag{5.58}$$

Combining these terms we get a description for the spectra:

$$H(\tau) = \frac{\{G(\tau) - G(\infty)\}}{\Lambda} \exp\left(-\frac{\Delta U(\tau)}{k_B T}\right) \tag{5.59}$$

where Λ is a normalisation factor to ensure the integral under the distribution is $G(\infty)$. The idea of a fluctuation in the configuration giving rise to a measure of the elastic moduli is analogous to Equation (5.31). However, we require a link between the distance moved and the timescale. This can be incorporated by using the long time self-diffusion coefficient. Unfortunately we do not have a good description of this for these systems. Assuming the long time self-diffusion coefficient to be a constant we know the mean square distance moved is proportional to the diffusion time:

$$\tau D_L = [R(\tau) - R]^2 \qquad (5.60)$$

so we can arbitrarily select a value of D_L and plot the data against τ/τ_r where τ_r occurs at the peak in the relaxation spectra. A typical spectrum is shown in Figure 5.19. The shape of the spectra is relatively insensitive to the magnitude of the pair potential between the particles so this common form applies reasonably well to a wide range of systems.

The relaxation function has been calculated and is compared with experimental data in Figure 5.16. The agreement between the model and the data is reasonable. The storage and loss moduli for a polystyrene latex have also been measured and compared to the model for the relaxation spectra. The data was gathered for a dispersion in 10^{-2}M sodium chloride at a volume fraction of 0.35 is shown in Figure 5.20.

The relaxation function has been used to predict the moduli and

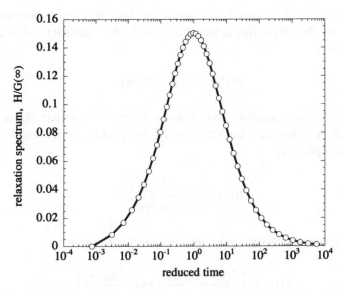

Figure 5.19   *A plot of a typical relaxation spectrum calculated from Equation (5.59)*

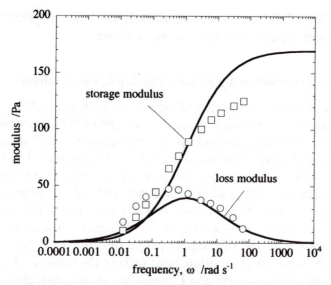

Figure 5.20 *A comparison of the storage and loss modulus obtained from an experiment for a polystyrene latex (points) with the spectra (Figure 5.19) calculated using Equation (5.59)*

agreement between the model and experiment is again quite reasonable. There is no doubt that this simplistic model could be improved from a direct simulation of the structure. Most of the restrictions in comparisons between theory and data however lie in the difficulty in producing systems which give a wide range of experimental responses. Often small amounts of solvent evaporation can result only in a minor change in volume fraction but a significant change in rheology. This permits only a very small range of concentrations where the relaxation time can be captured on the instrumental timescale. This restricted data set is a result of the nature of the material and the rate of change in rheological properties with concentration.

5.6 SIMPLE HOMOPOLYMER SYSTEMS

This section is primarily concerned with the behaviour of simple homopolymers. The development of viscoelastic theory was intimately linked with the study of polymeric species. This area of activity has led the way in the development of rheological models and experimental design and so is a very important area for the proto-rheologist to understand. So far in this chapter we have taken the approach of developing phase diagrams from a rheological perspective in order to understand linear viscoelastic behaviour. We will follow the same approach for simple polymers before broadening our argument to more complex systems.

5.6.1 Phase Behaviour and the Chain Overlap in Good Solvents

The linear viscoelastic properties of polymers, both as solutions and in the melt state, has been widely studied. The type of behaviour observed in the solution state is dependent upon concentration, which determines the region of the phase diagram occupied by the polymer. The boundaries to the zones in these phase diagrams is determined by the chain overlap parameter $c[\eta]$, and as such are defined in terms of the rheological properties of the system. The intrinsic viscosity here has the dimensions of reciprocal concentration. To illustrate our argument let us begin by considering the rheological properties of a rigid hard sphere dispersion. We know that for dilute hard spheres the intrinsic viscosity is defined in dimensionless form and Einstein has shown that, allowing for the hydrodynamic interactions between the solvent and the particle, it has a value of 2.5. We also know that for hard spheres we observe transitions from order to disorder at a volume fraction between 0.494 and 0.545. So we could define a parameter that is the product of the volume fraction and the intrinsic viscosity to demarcate where these transitions occur:

$$[\eta]\varphi = 2.5 \times 0.494 = 1.235 \qquad \text{and} \qquad [\eta]\varphi = 2.5 \times 0.545 = 1.363$$

In this section *alone*, in order to avoid confusion the intrinsic viscosity $[\eta]$ represents that of a particle and is dimensionless and the intrinsic viscosity $[\eta]$ represents that of a polymer and has units of $g\,cm^{-3}$. From these expressions we know that as we increase the volume fraction above each of these values, the product demarcates the structural transition. For particles there are many situations where it is relatively easy to determine the volume fraction, but for polymers the situation is more complex and it is most convenient to use a mass concentration. We can rewrite this relationship in terms of the mass of added *particles* in $g\,cm^{-3}$:

$$\varphi = \frac{4}{3}\pi a^3 \frac{c10^6 N_a}{M} \tag{5.61}$$

Here M is the mass of a mole of colloids, representing their molecular weight, and a is their radius. Now let us turn our attention to a polymer coil. The viscosity–concentration dependence of dilute polymers can be represented in a variety of forms. Each can relate to a particular molecular interpretation. The Huggins expression is a simple linearisation for homopolymers:

$$\frac{\left(\frac{\eta}{\eta_0} - 1\right)}{c} = [\eta] + ck_{\mathrm{H}}[\eta]^2 \tag{5.62}$$

where k_{H} is the Huggins coefficient. If we were to plot a graph of the specific viscosity, the left-hand side of the above equation, against concentration the intrinsic viscosity forms the intercept. It has the dimensions of $1/c$. Now let us suppose for the sake of illustration that our polymer coil has the hydrodynamic properties of a hard sphere. We can redefine our intrinsic viscosity as

$$[\eta]c = 2.5\varphi \tag{5.63}$$

substituting for the constants:

$$[\eta] = 2.5\frac{4}{3}\pi N_{\mathrm{a}}10^6\frac{R_g^3}{M} = 2.5\frac{4}{3}\pi N_{\mathrm{a}}10^6\frac{\bar{R}^3}{6^{3/2}M} \tag{5.64}$$

This simple hard sphere comparison illustrates the relationship between the root mean square length of a chain, its molecular weight and its intrinsic viscosity. For polymeric systems the hard sphere approach is too oversimplified and the relationship is expressed as

$$[\eta] = \Phi\frac{\bar{R}^3}{M} = 6^{3/2}\Phi\frac{R_g^3}{M} \tag{5.65}$$

where Φ is the Flory–Fox constant which is $\Phi \approx 2.5 \times 10^{29}$ when the molecular weight is given in Daltons and the lengths are given in metres. Equations (5.64) and (5.65) equate when $[\eta] \approx 1.46$. The value of $[\eta]c$ can be taken to indicate where phase changes in our system might occur. The intrinsic viscosity is a function of molecular weight so a plot of M versus c can be used to demarcate zones on the phase diagram. We can continue this concept a little further. We know that the square of the radius of gyration of our coil is proportional to the link length b and the number of links. In an ideal solvent:

$$R_g^2 \propto Nb^2 \tag{5.66}$$

So if we increase the molecular weight of the polymer then the number of links increases in direct proportion or

$$R_{g\theta}^2 = K_{R\theta}M \qquad (5.67)$$

where $K_{R\theta}$ is a constant. Substituting into our expression for intrinsic viscosity we obtain

$$[\eta]_\theta = K_\theta M^{1/2} \qquad (5.68)$$

where K_θ is a constant. The subscript θ indicates that this expression is for a polymer coil under θ conditions. The coil will expand or contract depending upon the solvency. In addition the internal degrees of freedom possessed by the chain reduces as the molecular weight reduces. We can write

$$[\eta] = KM^v \qquad (5.69)$$

$$R_g^2 = K_R M^{2v} \qquad (5.70)$$

where the exponent $v = 3/5$ is for a good solvent, and K and K_R are constants. Defining the boundaries for rigid hard spheres is less problematic than for deformable polymers which interpenetrate or collapse. For example with monodisperse hard spheres we know that once the volume fraction exceeds 0.74 the system becomes a solid as the spheres touch and are in a close packed structure. We may define a concentration φ^* as

$$\varphi^* = 0.74 = \frac{4}{3}\pi a^3 \rho^* \qquad (5.71)$$

This defines the number density ρ^* and hence concentration. This is determined by the particle radius. Analogous behaviour exists for polymers and there is a concentration at which polymer coils just touch:

$$\varphi_m = \frac{4}{3}\pi R_g^3 \frac{c^* N_a 10^4}{M} \qquad (5.72)$$

This definition of c^* is the same as that given in Section 2.4.2 with $\varphi_m = 1$ and different units. A variation in the units and symbols used is quite common in the literature and one should be wary of this. Combining Equations (5.65) and (5.72) we obtain an expression related to the overlap parameter:

$$c^* = \frac{1.08}{[\eta]} \qquad (5.73)$$

This demarcates the boundary between the dilute behaviour of a polymer and the semi-dilute regime. There can be slight variations in the value of the constant relating these properties, depending upon the assumptions used. Once the polymer is in the semi-dilute regime the coils overlap and interpenetrate. They do not necessarily form strong entanglements. In the region where the coil overlap begins, the expanding effect of a good solvent becomes screened by segments from neighbouring coils and the chains begin to collapse back toward their θ dimensions. This was investigated by de Gennes using scaling arguments and he suggested that the chain dimensions in the semi-dilute regime would reduce as

$$\bar{R}^2(c) = \bar{R}^2 \left(\frac{c^*}{c}\right)^{1/4} \tag{5.74}$$

A scaling argument does not describe the full functional form of the dependence of the chain dimensions on concentration. We would expect the dimensions of the polymer to be a smoothly changing function between c^* and the concentration $c^\dagger$ where the coils reach their limiting radius equivalent to the θ dimensions. However, to a first approximation we can suppose that Equation (5.73) applies up to the concentration $c^\dagger$:

$$\frac{c^\dagger}{c^*} = \left(\frac{\bar{R}^2}{\bar{R}^2(c^\dagger)}\right)^4 = \left(\frac{\bar{R}^2}{\bar{R}_\theta^2}\right)^4 \tag{5.75}$$

The ratio of the root mean square lengths is called the chain expansion factor:

$$\frac{c^\dagger}{c^*} = \alpha_e^8 \tag{5.76}$$

and represents the dimension of the chain relative to θ conditions. We begin by combining Equations (5.65), (5.67), (5.70) and (5.73) with (5.76) to obtain an expression for $c^\dagger$:

$$c^\dagger = \frac{1.08}{6^{3/2}\Phi} \frac{K_R^{5/2}}{K_{R\theta}^4} \tag{5.77}$$

which indicates that $c^\dagger$ does not depend upon molecular weight. However, this relies on the chain attaining its ideal configuration, *i.e.* $v = 3/5$, which is often not the case experimentally. Equation (5.77) can be difficult to evaluate directly. We can follow the argument of Graessley[25] and that of Berry, Nakayasu and Fox[26] to introduce the boundary between the semi-dilute and concentrated polymer behaviour. Graessley supposed that we can imagine that as the concentration is increased above c^* part of a polymer chain overlaps with another. There is a length of chain with a mean squared end-to-end length $\bar{R}_S^2$ characteristic of this screening distance, which depends upon the concentration but not on the molecular weight of the polymer. The screening length has a dimension that is equivalent to a molecular weight M_S of the same polymer in the same solvent at infinite dilution. Provided $M \gg M_S$ the mean chain dimensions $\bar{R}^2(c)$ can be taken as the number of steps M/M_S multiplied by the length of the step $\bar{R}_S^2$:

$$\bar{R}^2(c) = \frac{M}{M_S} \bar{R}_S^2 \qquad (5.78)$$

Expressing this in terms of the chain expansion factor:

$$\alpha_e^2(c) = \frac{\bar{R}^2(c)}{\bar{R}_\theta^2} = \left(\frac{\bar{R}_S^2}{M_S}\right) / \left(\frac{\bar{R}_\theta^2}{M}\right) \qquad (5.79)$$

and since $\bar{R}_\theta^2/M$ is independent of molecular weight we can write

$$\alpha_e^2(c) = \frac{\bar{R}^2(M_S)}{\bar{R}_\theta^2(M_S)} = \left(\frac{[\eta]}{[\eta]_\theta}\right)_{M_S}^{2/3} \qquad (5.80)$$

This means that the expansion factor depends only upon the end-to-end length $\bar{R}^2(M_S)$ of chains of molecular weight M_S relative to the chain in θ conditions. We can derive an effective concentration of the chains since this is related to the overlap concentration of the polymer of molecular weight M_S:

$$c = \frac{1.08}{[\eta]_{M_S}} \qquad (5.81)$$

Thus if we know $[\eta]$ and $[\eta]_\theta$ as a function of molecular weight we can plot the chain expansion factor as a function of concentration. A plot for polybutadiene from the work of Graessley is shown in Figure 5.21 and uses Equation (5.81) to describe the relationship between concentration and intrinsic viscosity.

At high concentrations the expansion factor approaches unity and this convergence can be used to define $c^\dagger$. When the concentration $c^\dagger$ is exceeded, the density of polymer segments becomes nearly uniform in solution. In this case we could estimate $c^\dagger$ as $0.11\,\text{g cm}^{-3}$. At concentra-

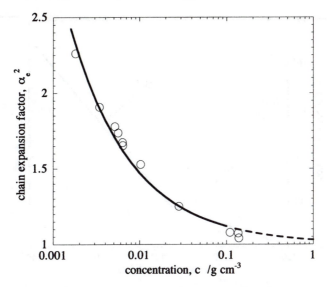

Figure 5.21 *A plot for the chain expansion factor for polybutadiene from the work of Graessley[25]*

tions above this simple mean field, arguments can be used to describe the rheological properties of the system. We would also like to define the boundary between interpenetration between the coils and where they entangle.

This will determine where the polymer chain dynamics are appreciably slowed and viscoelastic networks form. Experimental measurements for undiluted systems indicate that at low molecular weights the zero shear rate viscosity depends linearly on molecular weight when multiplied by a factor related to the frictional drag of the chain ends. At a critical molecular weight M_C the viscosity shows a greater dependence on molecular weight, being proportional to $M^{3.4}$. This is shown schematically in Figure 5.22.

This change has been associated with the onset of entanglements between the chains. These restrict the polymer motion and the relaxation time. Beuche[27] describes entanglement coupling in terms of the motion of one chain dragging another. This gave a molecular weight dependence of the viscosity of $M^{3.5}$. Whatever the details of the mechanism it is clear that the rate of change of viscosity with molecular weight is greater once entanglements occur. In reality, the change in viscosity for many systems is less sharp than shown in Figure 5.22, and different viscoelastic experiments can give different values for M_C. According to Graessley[25] and Doi and Edwards,[28] in the solution state at high concentrations of

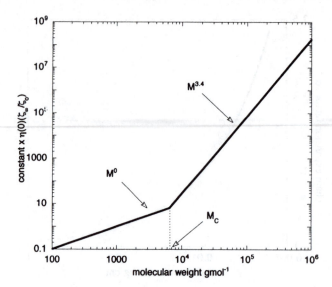

Figure 5.22 *A schematic of the log of the viscosity multiplied by the ratio of friction coefficients versus the log of the molecular weight*

polymer, the critical molecular weight between entanglements $(M_C)_{soln}$ can be approximated by

$$(M_C)_{soln} = \frac{M_C \rho_p}{c} \tag{5.82}$$

where ρ_p is the density of the polymer. So for a given polymer we can define our entanglement zones as:

$$cM > \rho_p M_C \quad \text{not entangled} \tag{5.83a}$$

$$cM < \rho_p M_C \quad \text{entangled} \tag{5.83b}$$

We can use Equation (5.73) to define the dilute to semi-dilute transition, Equation (5.80) to define the concentrated region and Equation (5.83) to define the onset of entanglements. An example of this is shown in Figure 5.23 for polybutadiene.

The various zones in Figure 5.23 define the areas where we expect different rheological responses. These transitions are not sharp and the actual location of the phase boundaries can differ slightly depending upon the assumptions used for defining the overlap parameters, or the presence of specificity between the solvent and the chain. In the following sections we will consider the rheological properties associated with some of these phases.

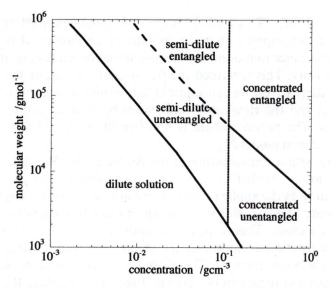

Figure 5.23 *Phase diagram for polybutadiene*

5.6.2 Dilute Solution Polymers[29]

At low concentrations, when uncharged polymers are dissolved in a solvent in which they do not crosslink or entangle, they possess a viscoelastic response through hydrodynamic and entropic effects. We can begin by considering an isolated chain in its quiescent state. The chain will be in constant motion. In the absence of any specific interactions, the chain will evolve to its maximum entropy state. We can represent the chain as N links or submolecules each with a length b. These links are formed from a few monomer units of the chain. The root mean square end-to-end length of the chain is

$$\langle \bar{R}^2 \rangle^{1/2} = b\sqrt{N} \tag{5.84}$$

Each submolecule will experience a frictional drag with the solvent represented by the frictional coefficient f_0. This drag is related to the frictional coefficient of the monomer unit ζ_0. If there are x monomer units per link then the frictional coefficient of a link is $x\zeta_0$. If we apply a step strain to the polymer chain it will deform and its entropy will fall. In order to attain its equilibrium conformation and maximum entropy the chain will rearrange itself by diffusion. The instantaneous elastic response can be thought of as being due to an 'entropic spring'. The drag on each submolecule can be treated in terms of the motion of the $N+1$ ends of the submolecules. We can think of these as 'beads' linked by entropic springs. This leads to a type of model described as 'bead–spring'. We can suppose that the motion of an individual bead has frictional resistance to motion in the solvent that depends only upon the solvent viscosity. This is termed the free draining model or the 'Rouse model'.[30] Alternatively we can suppose that beads on the same molecule can interact and the rheology is influenced by internal hydrodynamic interactions. The polymer chain is non-free draining and this is often termed the 'Zimm model'.[31]

We begin with a consideration of the Rouse model. We can visualise that there are a number of ways or modes (denoted by p) that the polymer can relax the applied strain. As we are considering a single chain we can ignore any viscoelastic contributions due to the centre of mass diffusion of a chain. This is the $p = 0$ mode. Higher modes involve the flexing and rotational motion of the chain each with a characteristic Maxwell relaxation time τ_p. There are as many modes as there are submolecules and these can be summed together to produce the relaxation behaviour of the chain. We can then multiply by the thermal energy and the number of chains per unit volume ρ_c to give the elastic moduli:

$$G(t) = \rho_c k_B T \sum_{p=1}^{N} e^{-t/\tau_p} \tag{5.85}$$

$$G'(\omega) = \rho_c k_B T \sum_{p=1}^{N} \frac{(\omega\tau_p)^2}{1 + (\omega\tau_p)^2} \tag{5.86}$$

$$G'(\omega) = \eta_0 \omega + \rho_c k_B T \sum_{p=1}^{N} \frac{\omega\tau_p}{1 + (\omega\tau_p)^2} \tag{5.87}$$

$$\eta'(\omega) = \eta_0 + \rho_c k_B T \sum_{p=1}^{N} \frac{\tau_p}{1 + (\omega\tau_p)^2} \tag{5.88}$$

where the relaxation times of each pth mode is given by

$$\tau_p = \frac{b^2 f_0}{24 k_B T \sin^2\{p\pi/2(N+1)\}} \tag{5.89}$$

The expression for the real component of the complex viscosity allows us to express the relaxation times as experimentally realisable parameters. In the low frequency limit we can rewrite Equation (5.88) in terms of the concentration c in $g\,cm^{-3}$:

$$\frac{\eta'(\omega) - \eta_0}{c} = \frac{N_a 10^6}{M} k_B T \sum_{p=1}^{N} \tau_p \qquad (5.90)$$

Inspection of Equation (5.68) indicates that in the dilute limit if we divide by the solvent viscosity we obtain the intrinsic viscosity:

$$[\eta] = \left(\frac{\eta}{\eta_0} - 1\right)/c = \frac{N_a 10^6}{M\eta_0} k_B T \sum_{p=1}^{N} \tau_p \qquad (5.91)$$

As the value of p increases the relaxation times become closer together and the contribution of large p over moderate frequencies becomes less important. So if we limit the modes we consider to $p < N/5$ we obtain a simpler expression for τ_p:

$$\tau_p = \frac{b^2 N^2 f_0}{6\pi^2 p^2 k_B T} \qquad (5.92)$$

This still contains parameters that are difficult to obtain experimentally, but we can rearrange Equation (5.91) as

$$[\eta] = \frac{N_a 10^6 k_B T \tau_1}{M} \sum_{p=1}^{N} \frac{\tau_p}{\tau_1} = \frac{N_a k_B T \tau_1}{M\eta_0} S_1 \qquad (5.93)$$

Now the sum $S_1 = \pi^2/6$ allows us to specify τ_p in terms of experimental parameters:

$$\tau_p = \frac{6[\eta] \times 10^{-6} \eta_0 M}{\pi^2 p^2 N_a k_B T} \qquad (5.94)$$

The Zimm model is a little more complex to evaluate but is essentially a sum of Maxwell models with a dependence on the sum of the modes as $p^{-3/2}$. This applies to all but the first few modes. A comparison between the two models is shown in Figure 5.24a and b.

At low frequencies the loss modulus is linear in frequency and the storage modulus is quadratic for both models. As the frequency exceeds the reciprocal of the relaxation time τ_1 the Rouse model approaches a square root dependence on frequency. The Zimm model varies as the $2/3^{rd}$ power in frequency. At high frequencies there is some experimental evidence that suggests the storage modulus reaches a plateau value. The loss modulus has a linear dependence on frequency with a slope controlled by the solvent viscosity. Hearst and Tschoegl[32] have both illustrated how a parameter h can be introduced into a bead spring

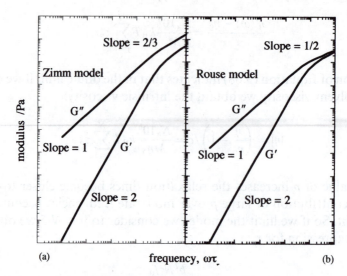

Figure 5.24   *The storage and loss moduli for (a) a Zimm[31] model and (b) a Rouse[30] model*

theory for viscoelasticity. This represents the hydrodynamic interactions so that a Rouse model is recovered when $h = 0$ and a Zimm model is recovered when $h = \infty$, showing that both approaches are limiting values of a spectrum of behaviour. We have assumed at the start a random walk for the polymer chain consistent with θ conditions. Thus the Rouse and Zimm models only apply in these limits. In practice experimental data on appropriate polymers rarely lies outside the range of values predicted by these two models and so h can be used as an adjustable parameter to reflect the solvency conditions. Other factors influencing chain conformation will in turn have an influence on the linear viscoelasticity. For example, the introduction of dissociating charged groups into the polymer chain will make the conformation of the chain dependent on pH and salt. The location and number of charges are important in determining the expansion of the chain. Experimental data on sodium polystyrene sulphonate and computer simulations suggest an increase in the value of the longest relaxation time due to a lengthening of the end-over-end rotation time.

An extreme case of tumbling dominating the relaxation behaviour can be seen with rigid rod-like species. The underlying concept is that the application of a strain is relaxed by the rotary diffusion of the rod. This gives rise to a single relaxation time. The precise shape of the molecule determines the form of the storage and loss moduli with frequency and there are a number of solutions. The effect of branching, molecular weight distributions and unusual polymer conformations can be predicted using the bead–spring model.

5.6.3 Undiluted and Concentrated Non-entangled Polymers

Above a concentration $c^\dagger$ the polymer can be considered to be concentrated, as shown by the phase diagram in Figure 5.23. At low molecular weights there is no significant entanglement coupling. As there is little or no solvent present, the hydrodynamic interactions can be considered to be negligible and the Rouse model is appropriate for describing the chain viscoelasticity. The density of the polymer controls the magnitude of the elasticity:

$$G(t) = \left(\frac{\rho_p N_a}{M}\right) k_B T \sum_{p=1}^{N} e^{-t/\tau_p} \tag{5.95}$$

$$G'(\omega) = \left(\frac{\rho_p N_a}{M}\right) k_B T \sum_{p=1}^{N} \frac{(\omega\tau_p)^2}{1 + (\omega\tau_p)^2} \tag{5.96}$$

$$G''(\omega) = \left(\frac{\rho_p N_a}{M}\right) k_B T \sum_{p=1}^{N} \frac{\omega\tau_p}{1 + (\omega\tau_p)^2} \tag{5.97}$$

However, in such a high concentration regime we can no longer represent the relaxation times (Equation (5.92)) in terms of the intrinsic viscosity. In the low frequency limit, because there is no permanent crosslinking present, the loss modulus divided by the frequency should equate with the low shear rate viscosity, or

$$\eta(0)\omega = G''(\omega) = \left(\frac{\rho_p N_a}{M}\right) k_B T \sum_{p=1}^{N} \omega\tau_p \tag{5.98}$$

If we perform the same substitutions made in Equations (5.93) and (5.94) we get

$$\tau_p = \frac{6\eta(0)M}{\pi^2 p^2 N_a k_B T} \tag{5.99}$$

This relationship between the relaxation modes and the low shear viscosity is an important one. It indicates that the longest Rouse relaxation time, *i.e.* the $p = 1$ mode:

$$\eta(0) = \frac{\tau_1 \pi^2 N_a k_B T}{6M} \tag{5.100}$$

determines the zero shear viscosity. This applies to both undiluted and highly concentrated polymers below their entanglement concentrations. We know that in this concentration region of the polymer the viscosity has a near linear dependence on molecular weight (Figure 5.22). From Equation (5.92) we can see that τ_1 depends on the square of the number of links in the chain, and so is equivalent to the square of the molecular weight. It follows then that the friction coefficient should be independent of molecular weight, at least at molecular weights less than M_C and its concentrated solution equivalent $(M_C)_{soln}$. However, experimentally this is not always observed, particularly at low molecular weights. The reason for this is thought to be the increasing importance of the ends of the polymer or 'tails'. These are less restrained than the monomer units in the centre of the chain and so occupy a larger free volume. The number of molecular tails per unit volume is proportional to M^{-1}. So if we allow for this the ratio of the low shear viscosity to the molecular weight $\eta(0)/M$ should tend to a constant value as the molecular weight is increased. This is where the monomeric friction coefficient is independent of molecular weight. We can define this value of the monomeric friction coefficient as ζ_∞ and the associated fractional free volume at this molecular weight as f_∞. This can be modelled in terms of the free volume argument proposed by Doolittle for viscosity (Section 3.3.4). In a liquid diffusion is possible when the free volume exceeds a critical value and this diffusive motion determines the friction coefficient. The same idea can be applied to the polymer 'tails'. It takes the form of an activation 'volume':

$$\zeta_0 = \zeta_\infty \exp\left(B\frac{f_\infty - f_M}{f_M f_\infty} \right) \qquad (5.101)$$

$$f_M = f_\infty + \frac{A}{M} \qquad (5.102)$$

where f_M is the fractional free volume at molecular weight M, B is a constant close to unity and A is a constant that relates the molecular weight to the free volume. So if $\log_{10}[\eta(0)\zeta_\infty/\zeta_0]$ is plotted against $\log_{10} M$ a slope of unity is obtained below the critical molecular weight for entanglements. This relationship has been convincingly demonstrated by Berry and Fox with data following Figure 5.22 for a wide range of systems. This indicates the role of the molecular weight on the free volume and the transition to the dependence $M^{3.4}$ when the polymers entangle.

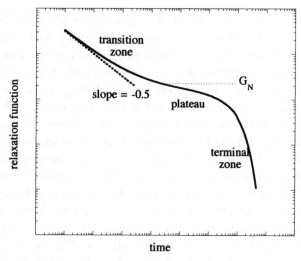

Figure 5.25 *Relaxation zones*

5.6.4 Entanglement Coupling

The rheological properties of simple polymers can show a change in slope with an increase in the concentration or molecular weight of a system. This is observed in both concentrated and semi-dilute regimes and is thought to be due to entanglements. The stress relaxation behaviour of the polymer shows a marked change from that of the Rouse model. A schematic is shown in Figure 5.25. The rapid decay of the relaxation function is determined by the longest relaxation time and this is called the *terminal zone*. At shorter relaxation times there is a plateau region associated with the elasticity due to entanglements G_N. At still shorter times the relaxation function increases and this is called the *transition zone*. Functions such as those described in Section 4.8.3 can be used to describe behaviour in this region.

The development of a second relaxation process is characteristic of entanglement formation. This additional relaxation process is accompanied by a sharp change in viscosity (highlighted in Figure 5.23) with molecular weight, and is also seen with concentration. The Rouse and Zimm models are concerned with the viscoelastic behaviour of a chain relaxing independently of the others around it. The presence of longer relaxation time processes indicates a slowing of the chain motion and thus can be considered to be due to the interactions between chains. In the absence of strong specific interactions between elements of the chain the slowing can be considered as due to mutually imposed topological constraints of the chain motion. The chains are becoming entangled and looped around each other.

Although this is an old concept the exact nature of an entanglement is still a matter for healthy debate. Relatively bulky molecules such as hydroxyethyl cellulose can show entanglement phenomena as readily as polystyrene or polyisoprene, despite each molecule possessing different molecular architecture. Ultimately in modelling these systems we attempt to reduce the chain to some simple properties such as a link length and a friction coefficient, hoping to collapse all the complex chemistry into these types of parameters. This means we will lose some of the details of the chemistry of the polymer. Large bulky chains require more monomer units to entangle and several monomer units will be required for an equivalent flexible link. For a given polymer one can envisage a number of entanglement sites occurring. There will be an average length and molecular weight between these sites. This is called the molecular weight between entanglements, M_e. This mass must be less than the total mass of the chain. At the site of the entanglement we can visualise that there are a range of timescales which are shorter than the time required for this entanglement to move and so relax the applied strain. The elasticity due to entanglements is thus given by the ability of the chain to store elastic energy between these sites. Suppose our site is tetrafunctional, $i.e.$ it is composed of two chains crossing. There can be more than one site per chain. The number of sites is determined by M/M_e. For each chain there are two tails, so the number of chains that are linked by entanglement sites is $(M/M_e) - 2$. We can now calculate the number of elastically effective chains ρ_e to give us the plateau modulus. For large values of M/M_e the plateau modulus is given by

$$G_N = \rho_e k_B T \approx \rho_c \left(\frac{M}{M_e}\right) k_B T \qquad (5.103)$$

This model, whilst not entirely satisfactory, can give a reasonable prediction of the elasticity in the plateau zone. The approach does not readily lend itself to a prediction of the dynamic behaviour. There have been numerous attempts to predict the low shear viscosity using entanglement models. For example Graessley obtained

$$\eta(0) \approx 2.7 \times 10^{-3} \frac{\rho_p N_a \zeta_0 b^2 M^{3.5}}{M_e^{2.5} x m_0^2} \qquad (5.104)$$

where m_0 is the mass of a monomer unit. Whilst it has been possible to predict the molecular weight dependence in this region with only a small error, a general treatment for the viscosity of semi-dilute and concentrated entangled polymers still remains problematic. It is notable that both the viscosity and the elasticity are controlled by the molecular weight between entanglements. A prediction of this for different polymers would be very useful for gauging the relative magnitude of the rheological responses. Beuche has suggested that the critical molecular weight is related to the molecular weight between entanglements as roughly $M_C \approx 2M_e$.[27] Ferry has tabulated available experimental data on M_e.[29] It is difficult to draw an overall conclusion from this data for M_e. The typical range of values can be estimated from the following expression:

$$j\frac{M_e}{m_0} = 100 \text{ to } 300 \tag{5.105}$$

where there are j atoms forming the chain per monomer unit. These models apply in the melt state. For a semi-dilute and concentrated polymer, as the concentration of the polymer is increased the modulus shows a non-linear dependence on concentration, *i.e.*

$$G_N = \rho_c \left(\frac{M}{(M_e)_{\text{soln}}}\right) k_B T + K_c \rho_c^2 \tag{5.106}$$

where $(M_e)_{\text{soln}}$ is the molecular weight between entanglements in the solvent. The viscosity dependence on concentration also changes. This behaviour is not readily developed from a generalised treatment of entanglements but linear and non-linear responses have been predicted by reptation theory.

5.6.5 Reptation and Linear Viscoelasticity

The underlying idea behind entanglements enables the construction of various models to describe elements of linear and non-linear viscoelasticity. It is difficult to formulate these ideas into a fully coherent description of polymer dynamics. A major lateral step was taken when de Gennes[34] applied scaling concepts to polymers. We can illustrate this idea by closely following the argument given by Doi and Edwards.[28] The properties of a Gaussian chain are such that its dimensions and statistical distribution of the segments do not depend upon the length of an individual segment for a large number N. Suppose we reduce the number of segments by a factor λ, then the new number of segments is $N' = N/\lambda$. If the chain is the same length, it is clear from Equation (5.84) that the link length must change as $b\sqrt{\lambda}$. We can represent this transformation from the old chain to the new chain as

$$N \rightarrow N/\lambda \quad \text{and} \quad b \rightarrow b\sqrt{\lambda} \qquad (5.107)$$

If we knew how a physical property changed as we altered our chain we could deduce how this property depended upon b and N. There are several statistical measures of length of the chain, for example the radius of gyration:

$$R_g = \frac{b\sqrt{N}}{\sqrt{6}} \qquad (5.108)$$

or the root mean square end-to-end length:

$$\bar{R} = b\sqrt{N} \qquad (5.109)$$

We could say quite generally that the average dimension of a chain is linear in the link length but is some function of N, $f(N)$:

$$\text{average dimension} = f(N)b \qquad (5.110)$$

So if we apply the transformation in Equation (5.107):

$$f(N)b = f\left(\frac{N}{\lambda}\right)b\sqrt{\lambda} \qquad (5.111)$$

This can only be true when

$$f(N)b = \text{constant} \times b\sqrt{N} \qquad (5.112)$$

This transformation applies equally well to a non-Gaussian chain, for example in a good solvent where $v = 0.5$:

$$N \rightarrow N/\lambda \quad \text{and} \quad b \rightarrow b\lambda^v \qquad (5.113)$$

So Equation (5.112) becomes

$$\text{average dimension} = \text{constant} \times bN^v \qquad (5.114)$$

The constant in Equation (5.112) cannot be readily evaluated using scaling theory. Our transformation applies equally well to the radius of gyration or the root mean square end-to-end length, only the numerical constant changes. We would like to be able to apply this idea to the role of concentration in semi-dilute and concentrated polymer regimes. In order to do this we need to define a new parameter s, the number of links or segments per unit volume:

$$\rho_c = \frac{s}{N} \tag{5.115}$$

We can define a critical segment concentration s^* equivalent to ρ_c^* provided we allow for the total number of links. If we ignore the constants in the expression we obtain:

$$s^* = \rho_c^* N = \frac{N}{R_g^3} \frac{N}{b^3 N^{3\nu}} = b^{-3} N^{1-3\nu} \tag{5.116}$$

We can express our transformation in the semi-dilute and concentrated regime as given by Equation (5.113) but with concentration included. So we obtain

$$N \rightarrow N/\lambda \quad b \rightarrow b\lambda^\nu \quad s \rightarrow s/\lambda \tag{5.117}$$

From Equation (5.106) for the elasticity in the plateau region we could express the moduli in terms of some function of chains:

$$G_N = \rho_c k_B T f(\rho_c, N) \tag{5.118}$$

This function must be dimensionless so applying Occum's razor to the parameters the simplest function is given by the expression:

$$G_N = \frac{s}{N} k_B T f(b^3 s, N) \tag{5.119}$$

Now let us apply our transformation to the expression:

$$\frac{s}{N} k_B T f(b^3 s, N) = \frac{s}{\lambda} \frac{\lambda}{N} k_B T f(b^3 s \lambda^{3\nu-1}, N/\lambda) \tag{5.120}$$

This must be invariant, and this is only true when

$$G_N = \frac{s}{N} k_B T f(b^3 s N^{3\nu-1}) \tag{5.121}$$

Now a substitution can be made from Equation (5.116) for a term in the brackets:

$$G_N = \frac{s}{N} k_B T f\left(\frac{s}{s^*}\right) \tag{5.122}$$

A similar expression can be derived for viscosity:

$$\eta(0) = \eta_0 f\left(\frac{s}{s^*}\right) \tag{5.123}$$

The function f is not necessarily the same in each case – it is just used to denote the existence of a function mapping concentration to a rheological measure. In order for us to be able to use Equations (5.122) and (5.123) we need to describe each functionality. We can do this by employing a tube model to describe the chain dynamics. Consider a polymer chain entangled in a mix of other chains. It follows a tortuous path through the gaps formed by its nearest neighbours (Figure 5.26). The chain and the surrounding network is fluctuating with time, all the chains constantly in motion. Suppose we freeze the motion to form a mesh. We can represent the surrounding chains by points so that the path of the chain appears to be avoiding these obstacles.

There is a gap between the chain and its neighbours and so the chain has freedom of movement within this gap. The chain appears to be constrained to a tube, with the surrounding chains determining the diameter of this tube. If we unfreeze the network we can imagine every chain constrained to its own tube and each tube fluctuating in time. The Brownian motion of the chain within the tube was considered by de Gennes.[34] The chain itself is folded within the tube as shown in

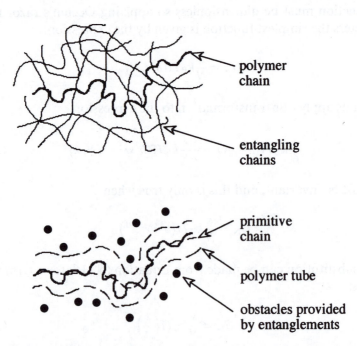

Figure 5.26 *Comparison between a chain in an entangled 'mesh' of polymer and the tube model*

Figure 5.26. Running along the centre of the tube is a primitive chain. This is the shortest path down the tube. The deviations of the polymer chain from this path can be considered as defects. The motion of these kinks or defects in the chain away from the primitive path allows the chain to move within the tube. The polymer creeps through the tube, losing its original constraints and gradually creating a new portion of tube. This reptilian-like motion of the chain was named by de Gennes from the Latin reptare, to creep, hence *reptation*.

The location of any individual tube changes with time but it has dimensions which are constant. If the tube contour is of length L we can define a parameter a, related to the mesh size of the polymer:

$$a = \frac{Nb^2}{L} \tag{5.124}$$

We can characterise the relaxation of the chain by two dominant relaxation processes: a short time process due to the fluctuations of the polymer within the tube, characterised by time τ_R; and a longer time process for the polymer to start to disengage itself from the tube τ_e. This process of tube disengagement is discussed further in Section 6.4.3. The characteristic time for the polymer to begin the process of escaping the constraints of a tube is given by

$$\tau_e = \frac{\zeta a^4}{k_B T b^2} \tag{5.125}$$

We can consider the friction coefficient ζ to be independent of the molecular weight. At times less than this or at a frequency greater than its reciprocal we expect the elasticity to have a frequency dependence similar to that of a Rouse chain in the high frequency limit. So for example for the storage modulus we get

$$G'(\omega) = G_N \left(\frac{\pi}{2}\omega\tau_e\right)^{1/2} \qquad \text{for } \omega\tau_e \geq 1 \tag{5.126}$$

The movement from the tube is restricted to odd-numbered Rouse modes, and this is determined by the nature of the motion required for a defect:

$$G'(\omega) = G_N \sum_{p,\text{odd}} \frac{8}{\pi^2} \frac{1}{p^2} \frac{(\omega\tau_d/p^2)^2}{1 + (\omega\tau_d/p^2)^2} \qquad \text{for } \omega\tau_e \geq 1 \tag{5.127}$$

A schematic of a representation of the storage modulus is shown in Figure 5.27 for the reptation model.

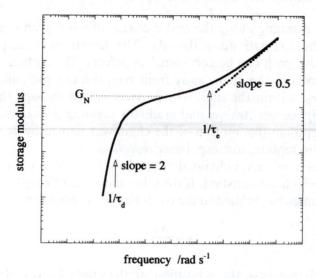

Figure 5.27 *The storage modulus predicted by the reptation model*

The plateau modulus is determined by the region in which these two relaxation modes cross. The plateau modulus, the low shear viscosity and the tube disengagement time are given in Section 6.4.3 as

$$G_N \cong \rho_c k_B T \left(\frac{Nl^2}{a^2}\right) = \frac{sb^2}{a^2} k_B T \tag{5.128}$$

$$\tau_d = \frac{\zeta N^3 b^2}{\pi^2 k_B T} \left(\frac{b}{a}\right)^2 \tag{5.129}$$

$$\eta(0) \cong \frac{\pi^2}{12} G_N \tau_d = s \frac{\zeta N^3 b^2}{12} \left(\frac{b}{a}\right)^4 \tag{5.130}$$

So far we have invoked reptation theory to describe the behaviour in the melt state. We can use scaling theory in the form of Equations (5.122) and (5.123) to express the concentration dependence of the modulus and viscosity. By inspection of Equations (5.128) and (5.130):

$$G_N \propto N^0 \quad \text{and} \quad \eta(0) \propto N^3 \tag{5.131}$$

As the functions in expressions (5.122) and (5.123) are dimensionless we could suppose that if we were to raise the ratios to some power, this would satisfy Equation (5.131). For these relationships to hold expressions (5.122) and (5.123) become

$$G_N = \frac{s}{N}k_B T f\left(\frac{s}{s^*}\right) = \frac{s}{N}k_B T\left(\frac{s}{s^*}\right)^m \propto N^0 \qquad (5.132)$$

$$\eta(0) = \eta_0 f\left(\frac{s}{s^*}\right) = \eta_0\left(\frac{s}{s^*}\right)^n \propto N^3 \qquad (5.133)$$

where n and m are defined by the proportionality. Now from Equation (5.116) we can state the value of the reciprocal of s^* and its dependence upon N. For this relationship to be maintained our power law is defined by substituting N^{1-3v} for s^*. This gives

$$G_N = \frac{s}{N}k_B T\left(\frac{s}{s^*}\right)^{1/(3v-1)} \propto M^0 \rho_c^{9/4} \qquad (5.134)$$

$$\eta(0) = \eta_0\left(\frac{s}{s^*}\right)^{3/(3v-1)} \propto M^3 \rho_c^{15/4} \qquad (5.135)$$

The proportionality is defined for $v = 3/5$. The molecular weight dependence of the viscosity is a little lower than is observed experimentally. Typically it depends on $M^{3.4}$. Graessley[35] noted that the predicted viscosity was also lower than the experimentally observed values, despite having a higher power law. He suggested the reason for this was that there is a broad transition in behaviour between 'pure' reptation and Rouse dynamics. Thus the theory and experimental data would converge at high molecular weights. However, simple calculations suggest that these molecular weights are not readily accessible by experiment. The functionality for the plateau modulus slightly overestimates the typically observed concentration dependence. This dependence tends to be closer to ρ_c^2 at higher concentrations. In equating scaling and reptation results we have assumed the tube dimensions are invariant with concentration, *i.e.* Equation (5.132) allows no dependence of the tube dimension a on N or the concentration ρ_c. We can investigate this by estimating the molecular weight between entanglements in solution polymers. By analogy to the melt situation we could assume:

$$(M_C)_{\text{soln}} \approx 2(M_e)_{\text{soln}} \qquad (5.136)$$

We can now define the solution case from the melt case from the value of M_C:

$$\frac{M_C \rho_p}{2c} \approx (M_e)_{\text{soln}} \qquad (5.137)$$

In Section 6.4.3 we will relate the tube dimensions to the molecular weight between entanglements for the melt. Making the appropriate substitutions from Equation (5.137) into Equation (6.100) we obtain

$$a^2 \approx 0.4Nb^2 \frac{Mc\rho_p}{Mc} \propto \frac{1}{\rho_c} \tag{5.138}$$

Substituting into Equation (5.128):

$$G_N = \left(\frac{m_L^2}{4 \times 10^5 Mc\rho_p N_a} \right) s^2 k_B T \propto M^2 \rho_c^2 \tag{5.139}$$

where m_L is the mass of a link or segment. This expression gives the required second power dependence, so providing in the solution state we allow the molecular weight between entanglements to vary with concentration we can obtain responses close to that seen in experiments. In conclusion we can see that reptation and scaling models do not solve all the problems in concentrated polymer solution rheology but they present a sound framework from which to develop our understanding.

5.7 POLYMER NETWORK STRUCTURES

The semi-dilute and concentrated polymer regimes are examples of the polymer chains showing connectivity across the entire system. They form an entangled network. As we have seen, the nature of the connectivity of the polymer chains has an enormous bearing on the rheological properties. We can think of entangled networks as 'physical networks' and they are essentially reversible networks upon, for example, dilution or variation in temperature. This reversibility may prove too long to be observed experimentally for very high molecular weight systems. The description of physical networks would include ions bridging from one chain to another, hydrophobic and hydrophilic (hydrogen) bonding, which are forms of transient bonding. The definition of 'chemical networks' is those connected structures that result from a chemical bond and are irreversible without cleavage of a chemical bond. A wide number of molecular and macromolecular systems can form networks and we shall now consider a few examples of these.

5.7.1 The Formation of Gels

One method of achieving a polymerised polymer-network is by the use of two distinct types of compounds, a monomer and a crosslinker. A simple example of this combination is styrene as a monomer and divinyl benzene as the crosslinker. The crosslinker, containing two vinyl units can link two poly(styrene) oligomers together whilst the chains themselves continue to grow. Depending upon the ratio of monomer to crosslinker, the polymers that form can be very lightly or heavily chemically crosslinked. This extent of crosslinking controls the rheology of the final product.

We can consider the formation of the polymer from a rheological point of view. Suppose we arrange conditions such that the final product is essentially a solid. Prior to initiation the solution monomers show a low viscosity which is only a little different from the solvent itself. As the reaction is initiated oligomers form and crosslink sites develop. The excluded volume begins to increase and the viscosity starts to rise. We can think of this as a progressive increase in the loss modulus with time and the extent of the reaction. As the chains become longer there is an increased ability to store energy and the storage modulus also rises. Initially we would anticipate the loss modulus would be greater than the storage modulus. However, as the reaction progresses toward the formation of a solid and completion, the storage modulus must become greater than the loss modulus. At some intermediate time between these two states the storage and loss moduli equate and we can define this time as a *gel point*. When the reaction is complete, some time after the gel point, we have complete connectivity and the molecular weight is effectively infinite. In this case we have a chemical gel. A similar argument could be applied to a physical gel, where connectivity results from reversible bonding between elements in the network. However, in both cases, because we know that the storage and loss moduli are frequency dependent, the gelation time would appear to depend upon the frequency selected.

Chambon and Winter[36] investigated this phenomenon experimentally and theoretically. They studied poly(dimethylsiloxane) (PDMS) networks. They began with a prepolymer of divinyl-terminated PDMS which they reacted with tetrakis(dimethylsiloxy) silane. This was performed in the presence of a platinum-based catalyst. This system was based on earlier work of Valles and Macosko.[37] The system had a major advantage in that the catalyst could be 'instantly' poisoned by minute amounts of sulphur. The reaction could be terminated and the rheology measured. However, current rheometric techniques allow the application of multiple wave forms which can now be used to circumvent this step. The storage and loss moduli were followed with time and at points either side of the cross-over between the storage and loss moduli frequency sweeps were performed. Prior to and post gelation the samples showed a progressively increasing storage and loss modulus. At the gel point a power law dependence can be used to describe the behaviour over a wide frequency range (Section 4.8.2)

$$G'(\omega) = \frac{\pi}{2\Gamma(m)\sin(m\pi/2)} S\omega^m \qquad (5.140)$$

$$G''(\omega) = \frac{\pi}{2\Gamma(m)\cos(m\pi/2)} S\omega^m \qquad (5.141)$$

The term S represents the strength of the network. The power law exponent m was found to depend on the stochiometric ratio r of crosslinker to sites. When they were in balance, *i.e.* $r = 1$, then $m = 1/2$. From Equations (5.140) and (5.141) this is the only condition where $G'(\omega) = G''(\omega)$ over all frequencies where the power law equation applies. If the stochiometry was varied the gel point was frequency dependent. This was also found to be the case for poly(urethane) networks. A microstructural origin has been suggested by both Cates and Muthumkumar[38] in terms of a fractal cluster with dimension D (Section 6.3.5). The complex viscosity was found to depend as:

$$\eta^* \sim \omega^{-2/(D+2)} \qquad (5.142)$$

by Muthumkumar. Using the above expressions to calculate the complex viscosity and equating the powers we get a relationship between the fractal dimension and the index m:

$$D = \frac{2m}{1-m} \qquad (5.143)$$

This relationship indicates that for a stochiometric gel the fractal dimension is 2. This argument need not be restricted to polymer gels but is valid for any colloidal fractal object. Another interesting feature to emerge from this modelling was the value of S, which did not reach a maximum value at $r = 1$ but reaches a plateau as $r \to 2$. However, the fractal dimension changes little in this region giving a value of $D = 2$; D only rises when the stochiometry reduces to $r < 1$. The network strength increases past the gel point and the viscosity begins to diverge towards infinity. The final network strength is an important reason for crosslinking polymers and this is considered further in the following section.

5.7.2 Chemical Networks

The crosslink density of a polymer network determines the number of elastically effective chains. Some of the chains are tied to a network and will not be able to relax the applied strain. This gives an elastic response at all frequencies and times. In a stress relaxation experiment the strain will decay due to some Rousian relaxation processes of all the chains, but some energy will be 'permanently' stored in the network. This long time relaxation process was modelled by the theory of rubber-like elasticity. The equilibrium modulus G_e can be expressed as

$$G_e = A_1 \left(\frac{\bar{R}_e^2}{\bar{R}^2} \right) v_n k_B T \tag{5.144}$$

where v_n is the number of moles of networked chains per unit volume. The mean squared end-to-end distance of the chain in the network is given by $\bar{R}_e^2$ and that of the same molecular weight of an unconstrained polymer is $\bar{R}^2$. The ratio of the terms is typically of the order of unity. The constant A_1 contains a number of factors including terms for the functionality of the link for more mobile crosslinks (Section 6.5.3). The variety and nature of the chemistry is critical in determining some of these constants and it is difficult to ascertain these without some experimental studies. Difunctional oligomers such as those discussed in the previous section, with controlled crosslinking stochiometry, are the systems most amenable to modelling. Differing degrees of cross-linking and changes in the functionality of a link site, *i.e.* the number of paths that meet at a crosslink, can have a major effect on the rheology. The polymers can become both entangled and crosslinked. The cross-link will restrict the motion of the entanglement and its reptative motion will become hindered, broadening the relaxation spectrum. For a simple network Mooney[39] suggested a model based on Rouse modes. The equations for stress relaxation and dynamic moduli can be expressed as

$$G(t) = v_n N_a k_B T \left(1 + \sum e^{-t/\tau_p} \right) \tag{5.145}$$

$$G'(\omega) = v_n N_a k_B T \left(1 + \sum \frac{(\omega \tau_p)^2}{1 + (\omega \tau_p)^2} \right) \tag{5.146}$$

$$G''(\omega) = v_n N_a k_B T \sum \frac{\omega \tau_p}{1 + (\omega \tau_p)^2} \tag{5.147}$$

These expressions are those for a viscoelastic solid and assume the 'front factor' is unity. The relaxation times, τ_p, are modified from simple Rouse values to allow for the motion of a strand rather than a complete chain.

We can define the Rouse relaxation time by Equation (5.92) multiplied by an additional factor for the shortened chains:

$$\tau_p = \frac{b^2 N \zeta_0 \rho_p^2}{6\pi^2 p^2 v^2 m_0^2 k_B T} \tag{5.148}$$

where N is the number of links in a chain *between* crosslinks. The approach here is a simple model of an ideal network without trapped entanglements. It represents the main features of a network formed from a concentrated polymer below the critical molecular weight for entanglements. The storage and loss moduli are shown in Figure 5.28. Polymer networks are best visualised by the complex compliance shown in Figure 5.29.

These were calculated using the relationships given in Section (4.6.1). The storage compliance starts from an initial compliance J_e given by

$$J_e = \frac{1}{G_e} \tag{5.149}$$

and reduces monotonically. The interesting feature is the peak in the loss compliance. This can be used as a characteristic of the network. The magnitude of the peak gives an indication of the network strength, and the peak position the characteristic relaxation time for the network.

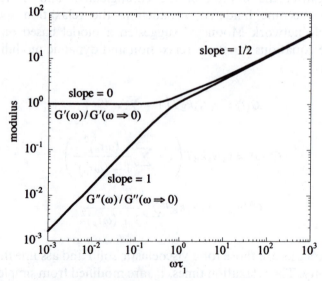

Figure 5.28 *The storage and loss moduli for a Mooney*[39] *model*

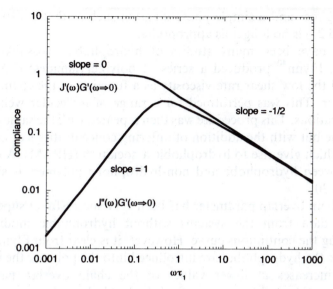

Figure 5.29 *The storage and loss compliance for a Mooney[39] model*

Other viscoelastic experiments can be performed but often the slow relaxation processes make it difficult to achieve an equilibrium response.

5.7.3 Physical Networks

It is possible to produce both chemical and physical networks in solution. One of the most important features of a physical network is the mobility of the crosslinking sites. For example in the non-linear regime, the applied stresses can be large enough to separate the transient links. Once the network structure is lost the sample readily undergoes shear thinning. Transient crosslinks can occur spontaneously if the polymer is synthesised with self-associating species. For polymers with a water-soluble backbone it is possible to add hydrophobic (alkyl) chains, which typically contain 10 to 16 carbons. In infinite dilution the polymer chains freely diffuse and give an intrinsic viscosity characteristic of their conformation in solution. However, as the concentration rises to a critical value the hydrophobic side chains on one polymer associate with those of the neighbouring chain. This is similar to the process of micellisation of surfactants (see Section 2.4.4). This network enhances the viscosity of the polymer. This is often seen as a rapid change in the viscosity as a function of concentration, measured by capillary viscometry. The important feature here is that the chain overlap parameter is no longer characteristic of the rheology transition between dilute and semi-dilute regimes and a simple phase diagram such as that constructed in Figure (5.23) is no longer as appropriate.

There have been many studies of hydrophobic crosslinking. For example, Flynn[40] produced a series of poly(acrylamides) (PAM) and recorded the low shear rate viscosity as a function of the chain overlap parameter. This was performed for a range of molecular weights and concentrations. This procedure was then repeated with the same polymer backbone but with the addition of differing concentrations of alkyl side chains which give rise to hydrophobic association (HPAM). A comparison between hydrophobe and non-hydrophobe polymers is shown in Figure 5.30.

The chain overlap parameter has been very successful at superimposing the data from the systems without hydrophobic modification, producing the continuous curve. However, it is clear from Flynn's work that once the hydrophobes are introduced into the polymer the viscosity rapidly increases at lower values of the chain overlap parameter. Increasing the mole percentage of hydrophobes also increases the viscosity at lower values of the chain overlap parameter. The position and number of the hydrophobes on a chain are important in determining the structure that forms and the onset of the increase in viscosity. The addition of side chains to hydroxyethyl cellulose modifies the network modulus as a function of concentration. This is discussed further in Section 2.3.4.

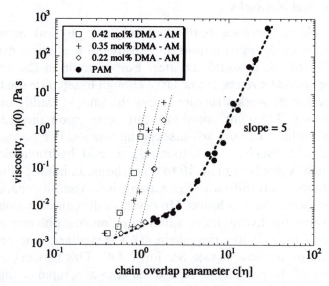

Figure 5.30   *The low shear viscosity versus chain overlap for polyacrylamide (PAM) and acrylamide dodecyl methacrylate copolymers (DMA–AM). The figures refer to the mole percentage of hydrophobic chains*

The situation is a little more complex on the addition of hydrophobes to the ends of the chains. Annabele and coworkers[41] studied hydrophobically-terminated polyethylene glycol. Each end was terminated by an alkyl group. They demonstrated that these can show narrow relaxation behaviour and tend to show a Newtonian viscosity at much higher concentrations than observed with systems with hydrophobic side chains. They do not follow reptation dynamics and it is the dynamics of the crosslink that is important in controlling the observed viscoelasticity. As the site of crosslinking is well known, the crosslink density can be predicted from the concentration of chains. This theoretical prediction overestimates the experimental observed network modulus, suggesting that the topology of the network is a strong function of concentration. Interestingly these studies showed that it was possible to superimpose storage and loss moduli data at different temperatures using a shift factor in frequency and modulus.

The notion of time–temperature superposition or reduced variables tends to work better for systems with lower solvent concentrations. The idea behind this is straightforward, for example for undiluted polymers examination of Equations (5.96) and (5.99) shows that the moduli and times have a simple temperature dependence:

$$G'(\omega) \propto T\rho_{\mathrm{p}} \quad \text{and} \quad \tau_p \propto \frac{\eta(0)}{T} \tag{5.150}$$

As we increase the temperature the relaxation time falls. So if we measure over the same frequency range at a temperature greater than some reference value we will measure higher frequency modes. The data superimpose if plotted relative to this reference temperature, so that we can define a function a_T such that

$$a_T = \frac{\eta(0)T_{\mathrm{ref}}\left(\rho_p\right)_{\mathrm{ref}}}{\eta(0)_{\mathrm{ref}}T\rho_p} \tag{5.151}$$

The subscript ref refers to the values at the reference temperature. If we plot the measurement frequency multiplied by the constant a_T all the data superimpose on the frequency axis. The correction for the moduli is to multiply by the function

$$\frac{T_{\mathrm{ref}}\left(\rho_p\right)_{\mathrm{ref}}}{T\rho_p} \tag{5.152}$$

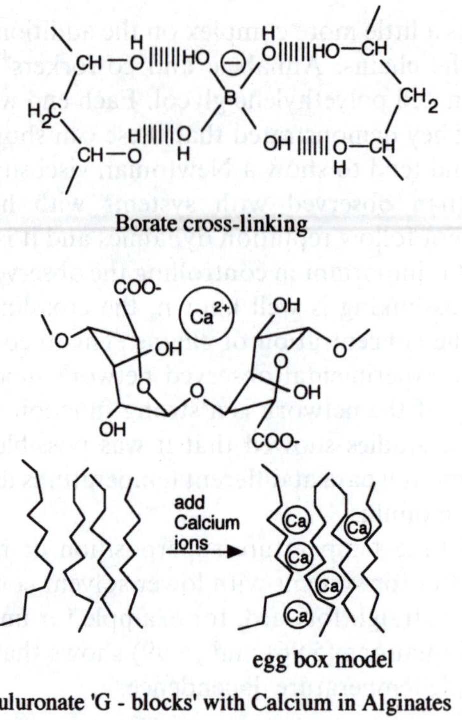

Borate cross-linking

Guluronate 'G - blocks' with Calcium in Alginates

Figure 5.31   *Transient crosslinks. The tetraborate ion in PVA and the calcium ion in an alginate*

Thus for undiluted polymers the relaxation behaviour can be examined over a wider range of apparent frequencies. Similar functions can be constructed for other regions of the phase diagram and other rheological experiments. The method of reduced variables has not been widely tested for aqueous crosslinked polymers. Typically these are polyelectrolytes crosslinked by ionic species. Some of these give rise to very simple relaxation behaviour. For example 98% hydrolysed poly(vinyl acetate) can be crosslinked by sodium tetraborate. The crosslink that forms is shown in Figure 5.31.

This results in narrow relaxation behaviour, as shown in Figure 5.32. The timescale of relaxation can be adjusted by altering the concentration of the tetraborate ion.

Ionic cross-linking can be achieved in a number of systems and different microstructures result. For rigid chains it is possible for the ion to coordinate in a specific manner with the chain. Alginates are a good example of this. These are natural polymers derived from a brown macro-algae. The method of extraction and the species of the algae influences the chemistry and hence the rheology of these polymers. There

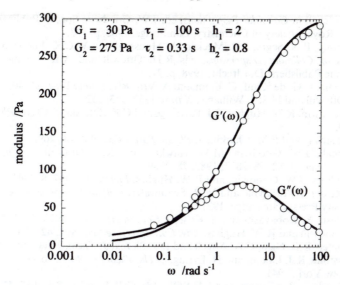

Figure 5.32 *Plot of the storage and loss moduli for a PVA gel (symbols). The data has been curve fitted using two log normal distributions (see Section 4.4.5)*

are two repeat groups in the chains, mannuronic (M) and guluronic (G) groups. The G groups are able to associate with divalent calcium. The level of G groups and the addition level of calcium determines the gel strength. The calcium ions is thought to be able to link chains giving rise to the so-called 'egg box' structure illustrated schematically in Figure 5.31. This is achieved readily in acid conditions. Naturally occurring polysaccharides can be crosslinked by a wide range of divalent ions, for example cobalt will crosslink hydroxypropyl guar. The specificity of the interactions means that a prediction of the rheological behaviour has to allow for the detailed chemistry of the polymer.

5.8 REFERENCES

1. D.A. McQuarrie, *Statistical Mechanics*, Harper and Row, New York, 1976.
2. B.J. Alder, and T.E.J. Wainwright, *J. Chem. Phys.* 1959, **27**, 1208; 1959, **31**, 459.
3. J.M. Kincaid and J.J. Weis, *Mol. Phys.* 1977, **34**, 931.
4. N.F. Carnahan and E.J.K. Starling, *J. Chem. Phys.* 1969, **51**, 635.
5. K.R.J. Hall, *J. Chem. Phys.* 1972, **57**, 2252.
6. L.V. Woodcock, *Ann. N. Y. Acad. Sci.* 1981, **371**, 274.
   P.N. Pusey and W. van Megen, in *Complex and Supramolecular Fluids*, eds S.A. Safran and N.A. Clark, Wiley-Interscience, New York, 1987, p. 673.
7. A.Vrij, J.W. Jansesn, J.K.G. Dhont, C. Parthmamanoharan, M.M. Kops-Werkhoven and H.M. Fijnaut, *Faraday Disuss.* 1983, **76**, 19.
8. I.M. Krieger and T.J. Dougherty, *Trans. Soc. Rheol.* 1959, **3**, 137.
9. N.W. Ashcroft and J. Lekener, *Phys. Rev.* 1966, **45**, 33.
10. J.K. Percus and G.J. Yevick, *Phys. Rev.* 1958, **110**, 1.

11. R.H. Ottewill in *Colloidal Dispersions*, ed. J.W. Goodwin, Special Publication No. 43, The Royal Society of Chemistry, Cambridge, 1982, p. 197.
12. P.N. Pusey, P.N. Segre, S.P. Meeker, A. Moussaid and W.C.K. Poon, in *Modern Aspects of Colloidal Dispersions*, eds R.H. Ottewill and A.R. Rennie, Kluwer Academic Publishers, Dordrecht, 1998, p. 77.
13. J. Mellema, C.G. de Kruif, C. Blom and A. Vrij, *Rheol. Acta* 1987, **26**, 40.
14. R.H. Ottewill and N.St J. Williams, *Nature* 1987, **325**, 232.
15. J.W. Goodwin, R.W. Hughes, S.J. Partridge and C.F. Zukoski, *J. Chem. Phys.* 1986, **85**, 559.
16. H.M. Lindsay and P.M. Chaikin, *J. Phys. (Paris) C3*, 1985, **46**, 269.
17. R. Buscall, J.W. Goodwin, M.W. Hawkins and R.H. Ottewill, *J. Chem. Soc., Faraday Trans.* 1982, **78**, 2873; 1982, **78**, 2889.
18. A. Bradbury, J.W. Goodwin, and R.W. Hughes, *Langmuir*, 1992, **8**, 2863.
19. W.B. Russel, D.A. Saville and W.R. Scholwater, *Colloidal Dispersions*, Cambridge University Press, Cambridge, 1991.
20. S. Hachisu, Y. Kobayashi and A. Kose, *J. Coll. Interface Sci.* 1973, **42**, 342.
21. J.W. Goodwin and R.W. Hughes, *Adv. Coll. Interface Sci.* 1992, **42**, 303.
22. R. Bohmer, K.L. Ngai, C.A. Angell and D.J. Plazek, *J. Chem. Phys.* 1993, **99**, 4201.
23. S. Glasstone, K.J. Laidler and H. Eyring, in *The Theory of Rate Processes*, McGraw-Hill, New York, 1941.
24. J.W. Goodwin, T. Gregory and J.A. Stile, *Adv. Coll. Interface Sci.* 1982, **17**, 185.
25. W.W. Graessley, *Polymer*, 1980, **21**, 258.
26. G.C. Berry, H. Nakayasu and T.G. Fox, *J. Polym. Sci., Polym Phys. Ed.* 1979, **17**, 1825.
27. F.J. Beuche, *J. Chem. Phys.* 1952, **20**, 1959; *Physical Properties of Polymers*, Interscience, New York, 1962.
28. M. Doi and S.F. Edwards, *The Theory of Polymer Dynamics*, Oxford University Press, Oxford, 1986.
29. J.D. Ferry, *Viscoelastic Properties of Polymers*, 3rd edn, Wiley, New York, 1980.
30. P.E. Rouse, *J. Chem. Phys.* 1953, **21**, 1272.
31. B.H. Zimm, *J. Chem. Phys.* 1956, **24**, 269.
32. J.E. Hearst, *J. Chem. Phys.* 1962, **37**, 2547.
    N.W. Tschoegl, *J. Chem. Phys.* 1963, **39**, 149.
33. M. Muthukumar and K. F. Freed, *Macromolecules*, 1978, **11**, 843.
34. P.-G. de Gennes, *Scaling Concepts in Polymer Physics*, Cornell University Press, Ithaca, NY, 1979.
35. W.W. Graessley, *J. Polym. Sci.* 1980, **18**, 27.
36. F. Chambon and H.H. Winter, *Polym. Bull.* 1985, **13**, 499.
37. E.M. Valles and C.W. Macosko, *Macromolecules*, 1979, **12**, 521.
38. M.E. Cates, *J. Phys. (Les Ulis, Fr.)* 1985, **46**, 1059.
    M.J. Muthumkumar, *J. Chem. Phys.* 1985, **83**, 3161.
39. M. Mooney, *J. Polym. Sci.* 1959, **34**, 599.
40. C.E. Flynn, Ph.D. thesis, University of Bristol, 1991.
41. T. Annabele, R. Buscall, R. Ettelaie and D. Whittlestone, *J. Rheol.* 1993, **37**, 695.

CHAPTER 6

Non-linear Responses

6.1 INTRODUCTION

The application of finite strains and stresses leads to a very wide range of responses. We have seen in Chapters 4 and 5 well-developed linear viscoelastic models, which were particularly important in the area of colloids and polymers, where unifying features are readily achievable in a manner not available to atomic fluids or solids. In Chapter 1 we introduced the Péclet number:

$$Pe = \frac{6\pi a^3 \sigma}{kT} \tag{6.1}$$

For a concentrated system this represents the ratio of the diffusive timescale of the quiescent microstructure to the convection under an applied deforming field. Note again that we are defining this in terms of the stress which is, of course, the product of the shear rate and the apparent viscosity (*i.e.* this includes the multibody interactions in the concentrated system). As the Péclet number exceeds unity the convection is dominating. This is achieved by increasing our stress or strain. This is the region in which our systems behave as non-linear materials, where simple combinations of Newtonian or Hookean models will never satisfactorily describe the behaviour. Part of the reason for this is that the flow field appreciably alters the microstructure and results in many-body interactions. The coupling between all these interactions becomes both philosophically and computationally very difficult.

Unfortunately the high Peclet number regime is where many rheological measurements are most easily made. High stresses and strain rates allow the development of simpler instrumental designs and lower sensitivities are required. It is also important to be aware of the fact that many applications require very high deformation regimes and it is important to understand these at least qualitatively. In practice much experimentally observed non-linear behaviour is difficult to describe in either a purely phenomenological or microstructural approach. Often it is dependent on very thorough physical and chemical characterisation. Even given this level of characterisation, models for describing the experimental behaviour become increasingly system-dependent and general rules are difficult to formalise. Having stated all the problems, this chapter is designed to address these difficulties. It uses simple models to give the reader a feel for possible origins of the responses you observe in your experiments. And remember your system is probably unique ...

6.2   THE PHENOMENOLOGICAL APPROACH

One of the great strengths of linear viscoelastic experiments is that the relationship between them is formally described. This has the advantage that if we have a thorough grasp of the ideas (not necessarily the underlying mathematics), it is relatively easy to predict what would happen as we apply different deformation regimes to the sample. Non-linearity can be more daunting. The Deborah number gives a good indication of the kind of test that is most appropriate. For systems which appear more elastic than fluid, *i.e.* $De > 1$, non-linear behaviour is easily recognised and simply characterised by the strain behaviour. For systems which appear more fluid than elastic, *i.e.* $De < 1$, non-linear behaviour is also easily recognised and simply characterised by the strain rate behaviour. In the following section we will consider simple models for the flow.

6.2.1   Flow Curves: Definitions and Equations

Most characterisation of non-linear responses of materials with $De < 1$ have concerned the application of a shear rate and the shear stress has been monitored. The ratio at any particular rate has defined the apparent viscosity. When these values are plotted against one another we produce flow curves. The reason for the popularity of this approach is partly historic and is related to the type of characterisation tool that was available when rheology was developing as a subject. As a consequence there are many expressions relating shear stress, viscosity and shear rate. There is also a plethora of interpretations for 'meaning' behind the parameters in the modelling equations. There are a number that are commonly used as phenomenological descriptions of the flow behaviour.

In some cases this approach, while having great practical utility, is a little unfair to the authors of the original equations. Certainly models such as those developed by Casson,[1] Cross,[2] and Woods and Krieger,[3] whilst widely used as phenomenological fitting equations, in fact all have an underlying microstructural interpretation. Others such as the Bingham model[4] have received some post rationalisation for the constants in the equation. If the viscosity reduces as the shear rate is increased a material is described as shear thinning, and if it increases it is described as shear thickening. We can class shear thinning flow curves by two sets of behaviour, plastic and pseudoplastic.

A *plastic* material is one which displays a yield stress. As the shear rate is reduced the apparent viscosity progressively increases, diverging and never reaching a constant value. If a stress is applied to a *plastic* material no flow will be observed until the stress exceeds the yield stress. Expressions describing *plastic* materials must include at least two terms, a yield stress and a limiting high shear viscous term. Expressions usually describe shear stress in terms of shear rate:

Bingham
$$\sigma = \sigma_B + \eta_{pl}\dot{\gamma} \tag{6.2}$$

Casson
$$\sigma = \left(\sqrt{\sigma_C} + \sqrt{\eta_{pl}\dot{\gamma}}\right)^2 \tag{6.3}$$

Herchel–Bulkley
$$\sigma = \sigma_{HB} + \left(\eta_{pl}\dot{\gamma}\right)^n \tag{6.4}$$

The term η_{pl} is the plastic viscosity and σ_B, σ_C and σ_{HB} are the Bingham, Casson and the Herchel–Bulkley yield stresses.

A *pseudoplastic* material is one which displays a limiting high shear rate viscosity. The stress increases in proportion to the shear rate. As the shear rate is reduced the viscosity begins to increase. In practice it can increase over many orders of magnitude of viscosity but eventually the rate at which the viscosity increases reduces and a constant viscosity is achieved, no matter how low the rate becomes. The constant high shear viscosity plateau $\eta(\infty)$ and low shear viscosity plateau $\eta(0)$ are sometimes called Newtonian plateaus because they are characteristic of Newtonian behaviour, albeit over a limited shear range. Expressions describing *pseudoplastic* materials usually include three terms, a critical stress, b, or rate, g, determining the transition from the low shear to high shear behaviour and the viscosity at the high and low limiting shears. Expressions usually describe apparent viscosity in terms of shear rate or shear stress:

Cross

$$\eta(\dot{\gamma}) = \eta(\infty) + \frac{\eta(0) - \eta(\infty)}{1 + (Pe/g\eta(\dot{\gamma}))^m} \qquad (6.5)$$

Krieger

$$\eta(\sigma) = \eta(\infty) + \frac{\eta(0) - \eta(\infty)}{1 + (bPe/6\pi)^n} \qquad (6.6)$$

where n and m are constants. These expressions are discussed further in Chapter 2 and later in this chapter. One feature worth noting is that it is possible to extrapolate to an 'apparent yield' stress from the plastic viscosity when plotting the data for the Krieger–Dougherty equation as shear stress versus shear rate. For particles of radius a and for $n = 1$ this is given by

$$\sigma_B = \left(\frac{\eta(0)}{\eta(\infty)} - 1 \right) \frac{kT}{ba^3} \qquad (6.7)$$

It is important when using the term 'yield stress' to distinguish between an extrapolated value, sometimes called the 'dynamic yield stress' and a true or 'static yield stress'. The latter can only be observed for plastic solids whilst the former is readily obtained with pseudoplastic liquids. In practical terms this can be critical in evaluating the performance of a material.

There is an expression that does not truly fit either class of behaviour, for power law fluids which can be expressed in terms of stress, rate or apparent viscosity with relative ease. They can describe *shear thickening or thinning* depending upon the sign of the power law index n:

$$\sigma = A_c \dot{\gamma}^n \quad \text{or} \quad \eta(\dot{\gamma}) = A_c \dot{\gamma}^{n-1} \text{ (Ostwald–De Waele)} \qquad (6.8)$$

where A_c is a constant. If m is positive, shear thickening is seen, and if n is negative shear thinning is seen. One note of practical interest here is when $n = 0$. A plot of the log of viscosity versus the log of shear rate will give a slope of -1. For a concentrated system this is often indicative of wall slip between the measuring geometry and the sample. Figure 6.1 shows examples of pseudoplastic and plastic behaviour as functions of stress and strain.

If we compare the shear stress against shear rate it can be practically difficult to distinguish between plastic and pseudoplastic behaviour. However when the same data is represented on a log–log plot clear differences emerge. The low shear viscosity begins to reduce the stress

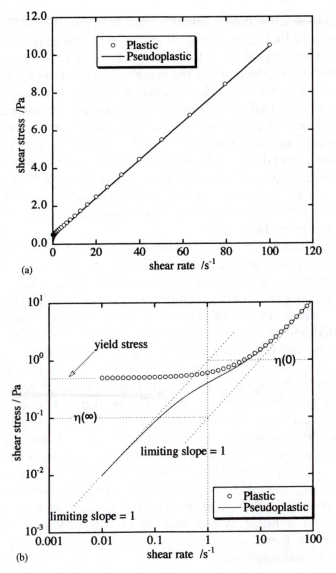

(a)

(b)

Figure 6.1 *Plot of shear stress versus shear rate for plastic and pseudoplastic materials. This is shown as (a) a linear–linear plot and (b) a log–log plot*

significantly. The application of a shear in the non-linear regime can give rise to forces normal to the shear direction. Flow instabilities can develop due to the viscoelastic behaviour of the samples at finite strains, to the extent where the sample cannot be constrained by the measuring geometry. Changes in the volume of material under deformation are commonly encountered. An example of this is when walking along a damp sandy beach and the sand around your footprint seems to dry as you walk. In a sense this is actually what happens. The deformation applied by your foot causes the sand grains to rearrange from the close packed structure produced by the water motion, and take up a greater volume, 'sucking in' the liquid in the process. The phenomena of volume changes under a shearing field is called *dilatancy*. The formation of structural order caused by deforming strains can result in some interesting effects. For example at high frequencies compression waves can be used to induce nucleation in supersaturated electrolyte solutions. For systems with slower diffusion dynamics than those of ions a gentler lower frequency oscillation can induce gelation. This effect is called *rheopexy*. This phenomena is time-dependent, and the shear equations in this section form constitutive equations which provide no information on this time-dependent behaviour.

There is a relationship that is used to cross between time and shear rate dependence regimes and that is the Cox–Merz rule.[5] The dynamic viscosity $\eta^*(\omega)$, when plotted as a function of frequency, has a similar form to a pseudoplastic curve although not necessarily displaying a high shear rate plateau. These observations apply for example to reasonably concentrated homopolymers. In the linear viscoelastic limit for a liquid the complex viscosity equates with the low shear rate viscosity, $\eta^*(\omega \rightarrow 0) = \eta(\dot{\gamma} \rightarrow 0)$. For some materials this rule can be extended into the non-linear response regime, equating the dynamic viscosity as a function of frequency, $\eta^*(\alpha\omega) = \eta(\dot{\gamma})$. The term α can be adopted to improve the agreement between the two sets of data. It would seem unlikely at first sight that an oscillating shear response would be comparable to a non-Newtonian finite strain response since the nature of the deformation is very different. However, good agreement is sometimes observed. It should be emphasised that in general, whilst the form of the experimental data can be similar $\eta^*(\omega)$ is a measurement that can be made with a linear system and as such is consistent with the mathematics of linear viscoelasticity, whereas shear thinning is very much a non-linear response.

6.2.2 Time Dependence in Flow and the Boltzmann Superposition Principle

The application of a shear rate to a linear viscoelastic liquid will cause the material to flow. The same will happen to a pseudoplastic material and to a plastic material once the yield stress has been exceeded. The stress that would result from the application of the shear rate would not necessarily be achieved instantaneously. The molecules or particles will undergo spatial rearrangements in an attempt to follow the applied flow field.

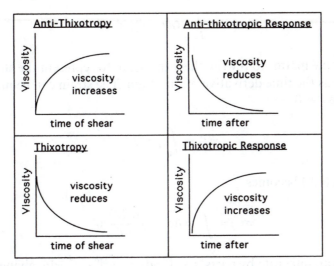

Figure 6.2 *Schematic of thixotropy and anti–thixotropy*

Simplistically one could visualise the molecules or particles in that sample either

- opposing the applied flow, resulting in an increase in viscosity with time at a constant shear rate – this is called *antithixotropy*
- following the flow field and rearranging to reduce energy dissipation resulting in a reducing viscosity with time at a constant shear rate – this is called *thixotropy*

These responses are shown diagrammatically in Figure 6.2. A Maxwell model is an example of a material in the *linear regime* that is antithixo-tropic, because the resistance to deformation increases as the spring extends until the maximum extension is reached. On cessation of flow the stress is relaxed and the viscosity falls. A thixotropic material has a viscosity that increases after cessation of flow.

Whilst the flow curves of materials have received widespread consideration, with the development of many models, the same cannot be said of the temporal changes seen with constant shear rate or stress. Moreover we could argue that after the apparent complexity of linear viscoeleastic systems the non-linear models developed above are very poor cousins. However, it is possible to introduce a little more phenomenological rigour by starting with the Boltzmann superposition integral given in Chapter 4, Equation (4.60). This represents the stress at time t for an applied strain history:

$$\sigma(t) = \int_{-\infty}^{t} G(t - t')\dot{\gamma}(t')dt' \tag{6.9}$$

It is a simple matter to rewrite this expression to represent the shear rate explicitly as the time derivative of the strain. The strain can be measured from time $t = 0$:

$$\gamma(t) = \int_{0}^{t} \dot{\gamma}(t')dt' \tag{6.10}$$

Equation (6.9) becomes

$$\sigma(t) = \int_{-\infty}^{t} G(t - t')\frac{d\gamma(t')}{dt'}dt' \tag{6.11}$$

Now we can integrate by parts, *i.e.* $\int_{a}^{b} udv = [uv]_{a}^{b} - \int_{a}^{b} vdu$, giving

$$\sigma(t) = [G(t - t')\gamma(t')]_{-\infty}^{t} - \int_{-\infty}^{t} \frac{dG(t - t')}{dt'}\gamma(t')dt' \tag{6.12}$$

This can be represented in terms of the memory function as

$$\sigma(t) = \int_{-\infty}^{t} m(t - t')\gamma(t, t')dt' \tag{6.13}$$

where $m(t)$ is the memory function, the time derivative of the relaxation modulus, and $\gamma(t,t')$ represents the strain at time t relative to that at time t' (Equation 4.62). Our sign convention on the memory function has been selected to be consistent with a model we shall adopt later in this chapter. Suppose we imagine a very simple stress relaxation function that depends on strain so that it is truly non-linear. The relationship could be quite simple. We could visualise a system with an invariant relaxation process and hence relaxation time.

A Maxwell model is a good candidate and we can incorporate the strain dependence by multiplying by a function $f(\gamma)$:

$$\sigma(t, \gamma) = Ge^{-t/\tau}f(\gamma) \tag{6.14}$$

This gives a strain-dependent relaxation function providing $f(\gamma) \neq \gamma$:

$$G(t, \gamma) = Ge^{-t/\tau}\frac{f(\gamma)}{\gamma} = G(\infty)e^{-t/\tau}\frac{1}{\gamma} = G(t)\frac{f(\gamma)}{\gamma} \tag{6.15}$$

We can introduce this function for the strain term into the Boltzmann superposition integral above. With the appropriate substitutions and change in limits we obtain an expression for the variation in stress with time:

$$\sigma(t) = G(t)f(\gamma) + \int_0^t m(t)f(\gamma)dt \qquad (6.16)$$

Now a shear rate represents the rate of change of strain so for the application of a constant shear rate to the sample the strain is the product of the rate and the time:

$$\sigma(t) = G(t)f(\dot{\gamma}t) + \int_0^t m(t)f(\dot{\gamma}t)dt \qquad (6.17)$$

This expression represents the stress at time t during an applied shear rate. If the shear rate is applied for long enough and the system is pseudoplastic the stress reaches a constant value. So extending the time toward infinity we get the limiting stress at any rate:

$$\sigma = \sigma(t \to \infty) = \int_0^\infty m(t)f(\dot{\gamma}t)dt \qquad (6.18)$$

The viscosity is shear stress divided by shear rate:

$$\eta(\dot{\gamma}) = \frac{1}{\dot{\gamma}} \int_0^\infty m(t)f(\dot{\gamma}t)dt \qquad (6.19)$$

Now we need to think carefully about the properties that this strain function might possess. Of course there are a wealth of possibilities so let us set the goal of describing pseudoplastic behaviour to give a material with a limiting viscosity at high and low rates. At first this appears a very daunting task but we can call on features from linear viscoelasticity. In the limit of zero strain and strain rate we would expect the material to have a low shear viscosity given by the linear viscoelastic expression in Chapter 4 (Equation 4.132):

$$\eta(0) = \int_{-\infty}^{+\infty} tG(t)\frac{f(\gamma)}{\gamma}d\ln t \bigg|_{\gamma \to 0} = G(\infty)\tau \qquad (6.20)$$

For this to happen we know that $f(\gamma) = \gamma$ in the low shear limit. As the shear stress is increased we also know that we want our viscosity to fall so we need to multiply our strain by a damping function that reduces from unity at low strains to a lesser value at high strains. A good candidate for this might be an exponential function in strain. The only problem with this choice would be that as the strain tended to infinity, $f(\gamma)$ would tend to zero and the viscosity would also equal zero. What we need is a constant high shear viscosity. So suppose we multiply the strain by a constant β which is less than unity. We can express Equation (6.20) in the high strain limit:

$$\eta(\infty) = \int_{-\infty}^{+\infty} tG(t)\frac{\beta\gamma}{\gamma}\mathrm{d}\ln t\bigg|_{\gamma\to\infty} = G(\infty)\tau\beta \tag{6.21}$$

Thus semi-intuitively we can write down our strain function:

$$f(\gamma) = \gamma e^{-(\alpha\gamma)^n} + \beta\gamma \tag{6.22}$$

The term α controls the critical strain where the system shear thins and n controls the rate of thinning. The low and high shear viscosity are given by

$$\eta(0) = G\tau(1 + \beta) = G(\infty)\tau \quad \text{and} \quad \eta(\infty) = G\tau\beta \tag{6.23}$$

Equations (6.17) and (6.19) allow us to describe the shear stress at any time t and at any shear rate. Ideally what we would like to achieve is to express $f(\gamma)$ in terms of a microstructural model based on the chemistry of our system. We shall see later that such an approach has been adopted by Doi and Edwards[6] for polymer solutions. The purpose of this phenomenological approach is to illustrate that constitutive equations like those of Krieger–Dougherty,[7] Cross and others can be obtained by a more formalised non-linear viscoelastic model. In fact the damping function in Equation (6.22) could be of any form that achieves the required reduction in strain. The integrals can be readily evaluated numerically but a simple exponential with $n = 1$ gives an analytical result. The steady state viscosity is given by

$$\eta(\dot{\gamma}) = \eta(\infty) + \frac{\eta(0) - \eta(\infty)}{(1 + \alpha\dot{\gamma}\tau)^2} \tag{6.24}$$

The stress at any time t also leads to a rather inelegant but analytical function:

$$\sigma(t) = G(t)f(\dot{\gamma}t) - I(t, \dot{\gamma}t) \tag{6.25a}$$

$$I(t, \dot{\gamma}t) = G\tau\dot{\gamma} \left\{ \begin{array}{l} -\beta - (1 + \alpha\dot{\gamma}\tau)^{-2} + \beta\left(1 + \dfrac{t}{\tau}\right)e^{-t/\tau} \\[2mm] + \left[(1 + \alpha\dot{\gamma}\tau)^{-2} - \dfrac{t}{\tau}(-1 - \alpha\dot{\gamma}\tau)^{-1}\right]e^{-\alpha\dot{\gamma}t - t/\tau} \end{array} \right\} \tag{6.25b}$$

We should be very wary over our choices for α and β since in the wrong combination they can give rise to physically unrealistic phenomena. This is a proviso that should be borne in mind when investigating many non-linear viscoelastic models. Some of the results of these non-linear viscoelastic calculations are shown in Figure 6.3. The three diagrams show the stress relaxation function as a function of both strain and time. The function is exponential in time but its magnitude reduces as the strain is increased. The pseudoplastic behaviour is shown and compared to both the Krieger–Dougherty and Cross models.

The most surprising result is that such simple non-linear relaxation behaviour can give rise to such complex behaviour of the stress with time. In Figure 6.3(b) there is a peak termed a *'stress overshoot'*. This illustrates that materials following very simple rules can show very complex behaviour. The sample modelled here, it could be argued, can show both thixotropic and anti-thixotropic behaviour. One of the most frequently made non-linear viscoelastic measurements is the *thixotropic loop*. This involves increasing the shear rate linearly with time to a given

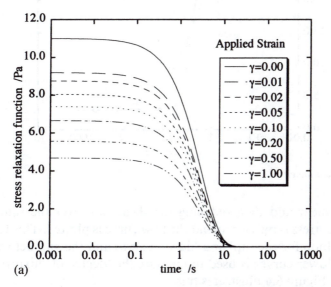

(a)

Figure 6.3 *Plot of a simple non-linear viscoelastic response for (a) the stress relaxation as a function of the applied strain, (b) stress as a function of time at a shear strain $\gamma = 1$ and (c) viscosity as a function of shear stress. ($\eta(0) = 33\,Pa\,s$, $\eta(\infty) = 3\,Pa\,s$, $\alpha = 1$, $\beta = 0.1$, $m = 0.35$ and $\tau = 1\,s$). Continued overleaf*

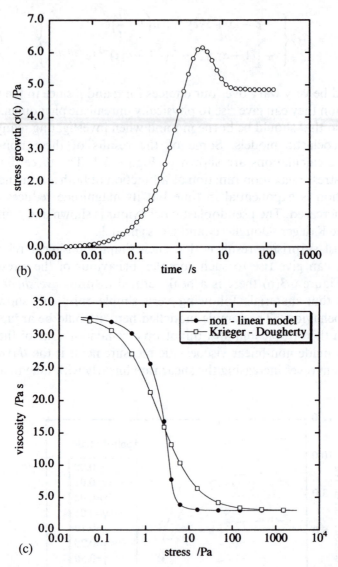

(b)

(c)

Figure 6.3 *(continued)*

maximum shear and then reducing the shear rate over the same time period. The stress is measured and the flow curve is plotted. This tends to generate a loop for the 'up' and 'down' curves and the gap between the upper and lower curves is used to measure the 'degree of thixotropy' of the system. Figure 6.4 illustrates this.

The problem with this classification is that it is very difficult to interpret. The non-linear approach used here is not ideal for developing loops because by decoupling the relaxation process from the strain they

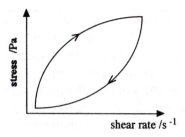

Figure 6.4 *Schematic of a thixotropic loop*

cannot allow for the recovery of the material. They do however give an indication of the behaviour qualitatively. The reason for the loops observed is associated with the form of the stress overshoot and the relaxation rate of the material. If the sample is deformed too quickly, steady state equilibrium is not achieved and we will sample a portion of the stress growth behaviour below the maximum stress, *i.e.* to the left of the peak shown in Figure 6.3(b). The increasing rate curve can appear below the decreasing one depending on the deformation time. However, if we deform the sample so we are at or just beyond the peak in the stress growth curve, the viscosity will show a relative reduction and the increasing rate curve will be above the reducing one. Clearly a crossover of shear up and down sweeps is possible if we were to cross from one side of the peak to the other during a sweep. The magnitude of the stress sweeping up relative to down depends upon the rate at which the structure breaks down relative to its ability to attain an equilibrium response. So at long experimental times the loop disappears and equilibrium has been achieved. The use of non-linear viscoelastic models has much to recommend it in that, as in the linear theory, they can be used to predict experimental responses. It provides justification for more empirical models of a seemingly arbitrary nature. However, they often contain a wealth of parameters that are difficult to predict but can be obtained by experimental fitting.

6.2.3 Yield Stress Sedimentation and Linearity

The non-linear response of plastic materials is more challenging in many respects than pseudoplastic materials. While some yield phenomena, such as that seen in clay dispersions of montmorillonite, can be catastrophic in nature and recover very rapidly, others such as polymer particle blends can yield slowly. Not all clay structures catastrophically thin. Clay platelets forming an elastic structure can be deformed by a finite strain such that they align with the deforming field. When the strain is released they do not necessarily recover to the initial structure; the microstructure of the material is 'permanently' changed. As a consequence both the yield and the nature of the viscoelasticity change. There is significant hysteresis. Modelling these changes with a phenomenological model has to account for what is effectively a rheologically induced phase change. Distinguishing between plastic and pseudoplastic samples can provide the rheologist with a severe challenge. One area where this is of great value is in the control of sedimentation in colloidal systems. If we consider a dispersion of particles denser than the surrounding medium, the particles will experience a force due to the action of gravity. For an isolated particle the sedimentation velocity is given by balancing the frictional drag of the solvent with the gravitational force. For spherical particles of radius a in a solvent with a viscosity η_0 we obtain

$$\Delta\rho \frac{4}{3}\pi a^3 g = v_0 6\pi\eta_0 a \qquad (6.26)$$

where $\Delta\rho$ is the density difference between the particles and the medium. The sedimentation velocity v_0 can be determined by simple rearrangement:

$$v_0 = \frac{2\Delta\rho g a^2}{9\eta_0} \qquad (6.27)$$

For a concentrated dispersion the particles feel the effects of all their neighbours and the sedimentation velocity can be represented semi-empirically as[8]

$$\frac{v}{v_0} = \left(1 - \frac{\varphi}{\varphi_m(\sigma)}\right)^{\beta\varphi_m(\sigma)} \qquad (6.28)$$

Where β is a constant. The packing fraction, $\varphi_m(\sigma)$, as we shall see later, may depend upon the stress applied, in this case a gravitational stress. In most commonly encountered circumstances the stress is low enough for the packing of the particles to be independent of the value. However, higher stresses can be applied during centrifugation and there will be an influence on this value relative to the **g** force used. This packing fraction represents the point at which the system becomes a viscoelastic solid and will not sediment. The above expressions do not describe the onset of sedimentation. A long induction time is possible before this will occur and this is very difficult to predict once a percolation threshold has been achieved. Our model supposes that our sample is pseudoplastic at a volume fraction below the maximum packing fraction. If it displays plastic behaviour the low shear viscosity will be infinite, although this does not necessarily mean the sample will not sediment. The gravitational stress acting on the particle depends upon its mass. For a large particle the local stress can be relatively large. If this were to exceed the yield stress the interparticle forces in the system would be unable to develop a structure which will support the particle and sedimentation can occur. The yield stress required to oppose sedimentation can be determined by projecting the gravitational force acting on the particle over an effective area. As a first approximation we can consider the suspension as an 'effective medium' which is acting on the surface of the particle, *i.e.* over an area $4\pi a^2$. We can now set up are our inequality for the yield stress and the gravitational stress:

$$\sigma_Y > \frac{4\pi\Delta\rho g a^3}{3.4\pi a^2} \approx \frac{\Delta\rho g a}{3} \qquad (6.29)$$

This expression should be used as an approximate rather than an exact formulation which would require interparticle forces and tensor analysis. The difference between a high zero shear rate viscosity and a significant yield stress is an important one. A dispersion with a high zero shear viscosity will sediment, albeit slowly, whereas a system with a yield stress exceeding that in Equation (6.29) will retain its integrity all the time that yield is maintained. Yielding phenomena can provide us with important information about storage stability. One approach to establishing a yield value is to apply an ever-increasing stress to a sample with time. The strain is monitored and once the strain increases appreciably the yield stress has been found. The major problem with relying on this technique lies in the speed at which the stress is increased. In the previous section we have seen how ramping the strain gives a multitude of different material behaviours. The same applies to linearly ramping the stress. If this experiment is performed quickly a small strain displacement will go undetected by the instrument and the instrument will increment on to the next stress. The yield stress will be overestimated. This is an important failing in sedimentation control because an overestimated yield could lead to a system which on paper seems stable but in reality separates. A gradual application of the stress is to be preferred over a more rapid sequence. Perhaps the best tests are those often employed in linear viscoelastic measurements. A sinusoidal oscillating stress or strain can be applied to the sample and storage and loss moduli can be measured. At low stresses or strains a frequency sweep will illustrate the presence of static modulus $G(0)$. This is indicative of the presence of a yield. This will be observed by the storage modulus failing to reach the baseline at low frequencies, the value of the modulus being displaced from the baseline by $G(0)$. However, a broad relaxation spectra or a Maxwell model with a long time relaxation process can also give rise to this observation so it is not unambiguous. Complementary to this test is the strain and stress sweep. This is a very good method of establishing linearity and its breakdown. In this test an oscillating strain (or stress) is applied to the sample. For a system in which the storage modulus is greater than the loss, normally the case in materials with a yield, as the strain is increased past the linear limit the storage modulus reduces. Ideally this test is best performed in the lower frequency limit to avoid the greater contribution from the high frequency elastic components. If a stress is applied the critical stress can be established directly from the curve. If a strain sweep is applied the apparent yield in the sample can be established by multiplying the critical strain by the value of $G^*(\omega)$ at that strain. A typical strain sweep for a charged colloidal dispersion is shown in Figure 6.5.[9]

A note of caution should be sounded here. Whilst the curves shown in Figure 6.5 are characteristic of many charged dispersions it should be recalled that once we apply a sinusoid to a non-linear system the response need not be a sinusoid. As the strain is increased into the non-linear region, the waveform passing through the sample becomes progressively distorted. The instrumental analysis in this case involves

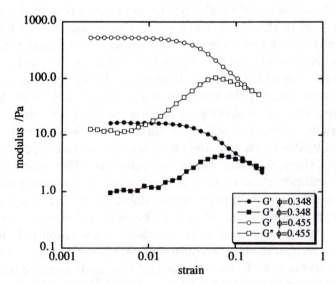

Figure 6.5 *Plot of the apparent storage and loss moduli as a function of strain for volume fractions of dispersions of silica particles (a = 150 nm, c = 10^{-3} M)*

applying a numerical Fourier transform to the signal and filtering to allow only components of the wave at the applied frequency to be analysed. Various harmonics can develop and this can provide further information on linearity. Inclusion of different terms and methods of analysis can change the form in the non-linear region. Without doubt the application and analysis of oscillating waves in the non-linear regime is the most difficult to model with rigorous viscoelastic equations. We have said that stress or strain controlled instrumentation can be used to obtain a test of linearity. Stress instrumentation can be designed to mimic strain rate responses. In order to achieve this, a stress controlled instrument must maintain a rapid and constant feedback between the torque applied and the subsequent displacement achieved in the measuring geometry. If the sample responds rapidly and can undergo changes more rapidly than the feedback and accompanying instrumental inertia allows, then the instrument will fail to maintain the correct strain. In the non-linear regime where the non-Maxwell processes can be very rapid, strain control can fail spectacularly. It is a useful design option but should be used, as with so many rheological measurements, with care.

In conclusion, a treatise on the phenomenological approach to non-linear responses is limited only by the amount of space available and the patience of both the author and the reader.[10] Simple equations for characterising data curves are useful tools. They allow us to reduce a complex set of data to a few parameters. Trends can often be more easily discerned from the parameters than from the original data. This is particularly the case for experiments where we are unable to hold all the chemical variables constant and only adjust one. The chemical engineer also finds simple models useful when solving heat and mass transfer problems. They can be used in finite element method calculations without the need for recourse to the molecular details underlying the system. This can on occasion provide cold comfort when a process fails and we need to establish the chemical origin. Then the connection must be made between simple models and non-linear flow behaviour. In the following sections this issue will be addressed.

6.3 THE MICROSTRUCTURAL APPROACH – PARTICLES

There is a wealth of microstructural models used for describing non-linear viscoelastic responses. Many of these relate the rheological properties to the interparticle forces and the bulk of these consider the action of continuous shear rate or stress. We will begin with a consideration of the simplest form of potential, a hard rigid sphere.

6.3.1 Flow in Hard Sphere Systems

In Chapter 3 (Section 3.5.2) the viscosity of a hard sphere model system was developed as a function of concentration. It was developed using an exact hydrodynamic solution developed by Einstein for the viscosity of dilute colloidal hard spheres dispersed in a solvent with a viscosity η_0. By using a mean field argument it is possible to show that the viscosity of a dispersion of hard spheres is given by

$$\eta(0) = \eta_0 \left(1 - \frac{\varphi}{\varphi_m(0)}\right)^{-[\eta]\varphi_m(0)} \quad \text{and} \quad \eta(\infty) = \eta_0 \left(1 - \frac{\varphi}{\varphi_m(0)}\right)^{-[\eta]\varphi_m(0)} \tag{6.30}$$

where $\varphi_m(\infty)$ and $\varphi_m(0)$ are the high shear and low shear packing limits of the dispersion. Practical realisation of such a system is difficult. Quasi-hard sphere models exist. Typical systems are formed from particles of polymethyl methacrylate coated by a layer of poly(12-hydroxystearic acid) or organophilic silica particles. Both systems are dispersed in non-aqueous media. When the system is placed under shear the viscosity reduces from the low to the high shear limit as the packing rearranges in flow. These systems are pseudoplastic but will display a Bingham yield stress [Equation (6.7)]. Currently the only reliable models that exist for describing the flow of these quasi-hard sphere systems with applied stress are empirical. We should recall that these models have developed on systems with an adsorbed layer so the volume fractions used are often defined from the flow properties of the particles or the diffusive behaviour. They rely on the use of hydrodynamic scaling to determine concentration in most instances. The Krieger equation (6.6) has proved a good description and constants can be derived for b from experimental data. There is of course uncertainty in the value of b. This can mainly be attributed to the difficulty of making the measurements at low stresses and the effect of slightly different deviations from hard sphere behaviour that different experimental systems possess. The same criticisms can be levelled at the low shear packing although at high shear rates layered flow seems to occur with $\varphi_m(\infty) = 0.605$. An example of experimental data demonstrating this is shown in Figure 6.6.

Intuitively we might associate the low shear limit with the order–disorder transition at $\varphi_m(0) = 0.495$. However literature data for the packing fraction in this limit is more widely scattered. We must remember that the approach to equilibrium in these systems can take a while to progress. So it is feasible that some systems have been measured away from the equilibrium state when the samples have been transferred and placed in the measuring geometry on an instrument. We could

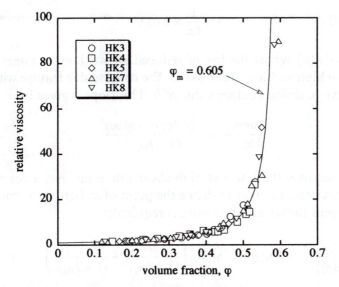

Figure 6.6 *The limiting high shear viscosity for quasi-hard sphere for PMMA particles in dodecane. (The particle has a different effective radii, HK3 = 419 nm, HK4 = 281 nm, HK5 = 184 nm, HK7 = 120 nm, HK8 = 162 nm.) The solid line is given by the Krieger equation (6.6) for a packing of $\varphi_m(\infty) = 0.605$*

speculate that the hard sphere viscosity profile can indeed be represented by Equation (6.6). A master curve for particle size and volume fraction can be developed by plotting η_n versus reduced stress σ_r:

$$\eta_n = \frac{\eta(\sigma) - \eta(\infty)}{\eta(0) - \eta(\infty)} = (1 + b\sigma_r)^{-1} \tag{6.31}$$

where in the low shear limit $\eta_n = 1$. To develop this curve we need to assign a value to b. We could develop a 'reference system' from a very simple argument. We know that when the Péclet number $Pe = 1$ the motion due to convection balances with the diffusive motion. We could use this to define a point on our flow curve but we need to make a decision about where this point occurs. There are several approaches. For example a plot of $\log(\eta_n)$ against $\log(\sigma_r)$ diverges to minus infinity with a slope given by

$$\eta_n = b\sigma_r^{-1} \tag{6.32}$$

If we extrapolate this line until it equates with the value of $\eta_n = 1$ where $\eta(\sigma) = \eta(0)$ then

$$b = \frac{1}{6\pi} \approx 0.053 \tag{6.33}$$

A plot of $\eta(\sigma)$ versus the log of reduced stress shows a linear slope between the high and low shear limits. We can use this feature with our master curve to define another value of b. This slope is given by

$$\frac{d\eta(\sigma_r)}{d\ln(\sigma_r)} = \frac{-b\sigma_r[\eta(0) - \eta(\infty)]}{(1 + b\sigma_r)^2} \tag{6.34}$$

The slope between the low and high shear rate limits has a constantly changing gradient. In order to define the point of inflection on this slope and its tangent the second derivative is required:

$$\frac{d^2\eta(\sigma_r)}{d\ln(\sigma_r)^2} = -b\sigma_r[\eta(0) - \eta(\infty)]\left(\frac{-2b\sigma_r}{(1 + b\sigma_r)^3} + \frac{1}{(1 + b\sigma_r)^2}\right) \tag{6.35}$$

The turning point occurs when the second derivative is zero, *i.e.* $b\sigma_r = 1$. The slope of the tangent is given by substituting this value into Equation (6.34). The equation for this line is

$$\text{tangent} = -\frac{[\eta(0) - \eta(\infty)]}{4}\ln\sigma_r + \eta(\infty)\left(\frac{1}{2} + \frac{\ln(b)}{4}\right) - \eta(0)\left(\frac{1}{2} - \frac{\ln(b)}{4}\right) \tag{6.36}$$

If we extrapolate this slope toward low rates, where the tangent equates with the low shear viscosity and assume the Péclet number here is unity we eventually obtain a value of $b = 2.55$. This defines the Péclet number as unity, at a stress somewhere just after the curvature of the viscosity curve deviates from the low shear limit. This seems quite an appropriate reference system. By setting the Péclet number to the appropriate value of b we can determine the variation in packing fraction with stress between the high and low shear limits to the viscosity:

$$\eta(\sigma) = \eta_0\left(1 - \frac{\varphi}{\varphi_m(\sigma)}\right)^{-[\eta]\varphi_m(\sigma)} \tag{6.37}$$

This defines a stress-dependent packing fraction. This is shown as a function of stress in Figure 6.7, which is sigmoidal in form, mimicking a mirror image of the viscosity profile.

As the shear rate increases so the particles begin to align with the flow field and pack more efficiently. This is a system dominated by the hydrodynamic forces overcoming the Brownian motion. Goddard, and

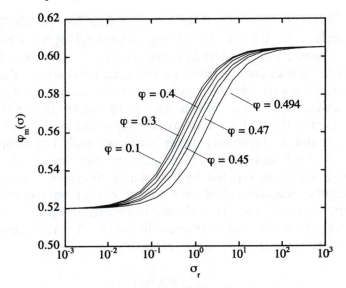

Figure 6.7 *Plot of the stress-dependent packing fraction $\varphi_m(\sigma)$ versus the reduced stress σ_r for maximum packing fractions of $\varphi_m(\infty) = 0.605$ and $\varphi_m(0) = 0.52$ and $b = 2.55$. This gives a relative viscosity of about 50 at the freezing transition*

Frankel and Acrivos[11] have obtained models with well-defined hydrodynamics for very high concentrations of rigid and elastic particles. Here the solvent forms thin films and we enter the region of lubrication theory. The expressions describing the flow do bear some similarities to the semi-empirical expressions developed at lower concentrations. For example Frankel and Acrivos give

$$\eta(0) = \eta_0 \frac{9}{8} \frac{(\varphi/\varphi_m)^{1/3}}{1 - (\varphi/\varphi_m)^{1/3}} \tag{6.38}$$

This is for a simple cubic lattice. As we include interparticle forces our ability to describe the system other than by numerical simulation becomes progressively more difficult to achieve. At present quasi-hard spheres at moderate to large volume fractions can only be modelled by analytical expressions that are empirical in origin. Simple models are available for other forms of interaction potentials.

6.3.2 The Addition of a Surface Layer

The expressions used for dispersions of quasi-hard spheres are usually measured on systems with a chemically grafted layer. This layer is chosen such that it is rigid and bristle-like in form. The idea is that this layer is relatively non-free draining and the hydrodynamic flow pattern around these particles is much like that with smooth hard rigid spheres. Suppose we have a system with a grafted or adsorbed layer that is less densely packed with various combinations of loops, tails and trains. Rigorous prediction of the role this layer may play on the local hydrodynamics requires a detailed structure of this layer to be known. In practice it is difficult to define this without extensive characterisation. Even given this the flow around the layer may prove computationally too complex to evaluate. A simple approach is to assume the particles are effectively larger by a thickness δ_H, the hydrodynamic thickness. This can be defined from a measurement of the viscosity. For a given model for the behaviour, we can define from the measurement an effective volume fraction, φ_{eff}, from the relative increase in volume. The relationship for spherical particles is given by simple geometry:

$$\varphi_{eff} = \varphi \left(\frac{a + \delta_H}{a} \right)^3 \tag{6.39}$$

For anisometric particles the ratio of the volume with adsorbed layer to that without provides a suitable approximation. The value of the apparent hydrodynamic thickness will depend upon a variety factors, for example for an adsorbed polyelectrolyte the extension of the layer will depend upon the mutual repulsion of dissociating groups on the backbone of the polymer. This gives us a layer thickness that is pH and electrolyte dependent. At very high concentrations particles with adsorbed flexible polymer layers can be concentrated to the level where the layers begin to interpenetrate. These systems display shear thickening. The main difficulty in establishing general rules for the onset of thickening is that it tends to occur at high particle concentrations. Here the separations between the core particles are small. This means that small differences in the surface structure between systems can prove difficult to characterise but are vitally important in controlling flow. Shear thickening can be caused by a number of factors. Clustering at high Péclet numbers has been suggested from computer simulations.[12] These probably represent the best prospect of understanding the behaviour. However, modelling of all but the best characterised systems is going to be extremely difficult.

6.3.3 Aggregation and Dispersion in Shear

The introduction of long-range interaction forces between colloidal particles can produce well dispersed or aggregated systems under quiescent conditions. The application of a shear field can induce changes in the structure. So for example when milling a dispersion, high energy shear fields are applied and this can overcome the interaction forces in a coagulum and disperse the system. The process is a complex balance of particle shape, interaction energy and the local turbulence. In the lower Reynolds number limits the application of simple shear fields can also overcome the forces holding flocs together or induce coagulation. Even in simple shearing fields a precise examination of this phenomena shows that it is difficult to model. However, it is possible to construct reasonably good maps of zones of stability and instability in systems of dilute to moderate concentration. The starting point is to examine the interaction forces between the colloidal particles. A typical interaction force curve is shown in Figure 6.8.

This force–distance curve is for the pair potential shown in Figure 5.9. This shows a steric barrier of 7–8 nm to prevent close approach of a pair of particles. A primary maximum can also prevent close approach of the particles under quiescent conditions and a secondary minimum will allow weak attraction between the particles. This will cause weak flocculation between the particles. Such a curve is shown in Figure 6.9a for a polystyrene particle with a 1 μm radius and $\kappa a = 100$ with a steric barrier of about 2 nm. When a shear field is applied the particles gain energy in the flow so collisions between the particles will be of higher energy. This gives the possibility of either disrupting the aggregates and

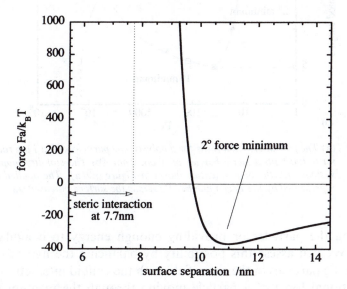

Figure 6.8 *The pair interaction force for the system with the pair potential shown in Figure 5.9*

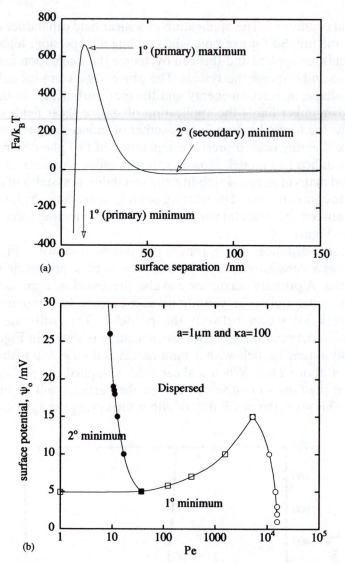

Figure 6.9   (a) *The pair interaction force for a polystyrene particle with a 1 μm radius and κa = 100 with a steric barrier of about 2 nm.* (b) *The stability map for the system with the pair potential shown in Figure 6.9(a). The applied shear is represented by the Péclet number Pe versus the surface potential ψ₀*

dispersing the particles or providing enough energy to coagulate the system. We can assess this possibility by balancing the hydrodynamic force on the particles in the shear field with the colloid interaction force. The frictional force of a particle moving through the medium with a velocity *v* is given by

$$F = 6\pi\eta av \qquad (6.40)$$

Now in convection we can obtain the velocity of the particle by multiplying the velocity gradient, *i.e.* shear rate, by the distance moved by the particle. We can think of this as some multiple of the particle radius, ξ:

$$F = 6\pi\eta\dot{\gamma}a^2\xi = 6\pi\sigma a^2\xi \qquad (6.41)$$

Suppose we normalise such that the force is dimensionless. This gives us the Péclet number:

$$\frac{Fa}{k_BT} = \bar{F} = Pe\xi \approx Pe \qquad (6.42)$$

As a reasonable approximation we can assume that one particle radii will provide a sufficiently large velocity to cause a change in structure. So if we equate a colloid interaction force with $\bar{F}$ we can construct the map in Figure 6.9b. This shows zones of aggregation relative to the applied shear field and the surface potential on the particles. Below Péclet numbers of 1, the system will be close to a quiescent state. A system can be considered stable against coagulation if the height of the primary maximum in energy is greater than $\sim 20k_BT$.

At very low surface potentials the attractive forces dominate and the system coagulates. For high Péclet numbers it is possible in principle to disperse particles by pulling them from the primary minimum. This might occur in milling. Here the hydrodynamic force is greater than the depth of the primary minimum. As the surface potential is increased the primary maximum grows to above $\sim 20k_BT$ and the system is stable against coagulation but can show secondary minimum flocculation, which is a weak flocculation. This is where, at close separations between the particles, an attractive energy exists which is greater than the thermal energy k_BT but where a primary maximum prevents coagulation. Theses systems are considered in more detail in the next section.

6.3.4 Weakly Flocculated Dispersions

One of the characteristics of weak flocculation is that the system is reversible. At low volume fractions the system will form some clusters and some single particles. The clusters can be easily disrupted by gentle shaking. As the concentration is increased the system will reach a 'percolation threshold'. The number of nearest neighbours around any test particle reaches 3 at about $\varphi = 0.25$ and the attractive forces between the particles lead to a connectivity throughout the system giving rise to a weak network. The application of a shear stress or shear rate gives rise to a system, which is dominantly thixotropic and shear thinning. As the break-up of the network progresses under shear, a steady state viscosity is achieved at a particular rate. If it is truly weak flocculation then at high shear rates we would expect the network to break up into individual particles. For spherical particles the Dougherty–Krieger expression can be applied. A starting point for modelling is to use the Bingham equation to 'fit' to the flow curve. Whilst the Bingham equation will not describe the steady state flow curve in detail the extrapolated Bingham yield stress can be related to the link energy and the rate at which links break. An analysis due to Michaels and Bolger[13] can be used to describe the breakdown process. We begin by calculating the energy dissipation rate:

$$\dot{E} = \frac{\mathrm{d}E}{\mathrm{d}t} = \sigma\dot{\gamma} = \eta_0 \left(1 - \frac{\varphi}{\varphi_{\mathrm{m}}(\infty)}\right)^{-[\eta]\varphi_{\mathrm{m}}(\infty)} \dot{\gamma}^2 + \sigma_{\mathrm{B}}\dot{\gamma} \qquad (6.43)$$

The Dougherty–Krieger term represents the energy dissipated per unit volume per second in order to maintain the hydrodynamic element of the flow in shear. The Bingham yield stress represents the energy dissipated in breaking the links per unit volume per second. This in turn is related to the energy of a link, *i.e.* the depth of the attractive minimum. As we are dealing with the steady state response of the system so the structure has reached a dynamic equilibrium, we know that under shear the rate at which links form equals the rate at which they break. So by calculating the rate at which the links form multiplied by the depth of the energy minimum we will have calculated the term on the right-hand side of Equation (6.43). We can argue that for every collision between particles in flow there is a probability, v, that a link will form. So if we can calculate the number of collisions and have an estimate for v the Bingham yield can be found. If we imagine a test particle and particles passing close to it in a dominant shear field a collision will occur when a particle centre passes closer than one diameter from the test particle centre. In order to determine the number of collisions we need to determine the particle flux J through a disc of diameter $4a$ placed around our test particle normal to the direction of the shear. This is shown schematically in Figure 6.10. The flux is the number of particles passing through the disc per second.

Let us first consider our disc around a test particle moving with velocity v in direction z in the flow. We can define an element of that disc as having a thickness $\mathrm{d}z$. The angle subtended by the ends of that segment we define as θ. The velocity is thus

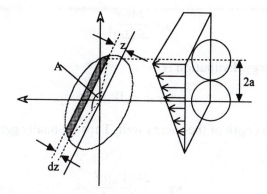

Figure 6.10 *The shear geometry for the collision*

$$v = z\dot{\gamma} = 2a\dot{\gamma}\cos(\theta/2) \tag{6.44}$$

We need to integrate this to obtain the volume. For a segment thickness
dz and width x,

$$dz = -a\sin(\theta/2)d\theta \quad \text{and} \quad x = 4a\sin(\theta/2) \tag{6.45}$$

Integrating to find the flux through the disc with ρ as the number density
of particles:

$$J = -2\rho \int_0^\pi -(2a)^3\dot{\gamma}\cos(\theta/2)\sin^2(\theta/2)d\theta \tag{6.46}$$

The total collision frequency is

$$c_p = \frac{vJ\rho}{2} \tag{6.47}$$

where the factor of 2 allows for the collision occurring between two
particles:

$$c_p = \frac{2}{3}v\rho^2(2a)^3\dot{\gamma} \tag{6.48}$$

Expressing this as volume fraction:

$$\varphi^2 = \left(\frac{\rho\pi(2a)^3}{6}\right)^2 \tag{6.49}$$

$$c_p = \frac{24v\varphi^2}{\pi^2(2a)^3}\dot{\gamma} \qquad\qquad (6.50)$$

The rate of energy dissipation is

$$\sigma_B\dot{\gamma} = vc_p|V_m| \qquad\qquad (6.51)$$

where V_m is the depth of the energy well. Thus we finally get the Bingham yield stress as

$$\sigma_B = \frac{24v|V_m|\varphi^2}{\pi^2(2a)^3} \qquad\qquad (6.52)$$

The value of v has been shown by hydrodynamic calculations to be fairly close to unity. This simple model was developed by Michaels and Bolger[13] and includes the main features of importance in controlling the Bingham yield stress in weakly attractive systems. Weak flocculation (Section 5.4) can be achieved, for example, by adsorbing a thin surfactant layer onto a polystyrene latex and by adjusting the level of electrolyte. Data for such a system is shown in Figures 5.9 and 6.11. The depth of the energy well in this case about $-7k_BT$ and the attractive force minimum $370k_BT/a$.

The attractive energy between the particles can be calculated by plotting σ_B versus φ^2. The first feature to note is the change in slope at

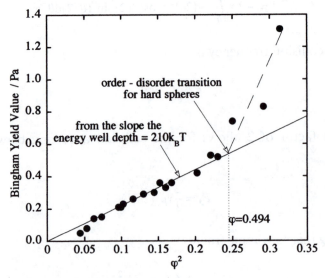

Figure 6.11 *The Bingham yield stress as a function of volume fraction for the system with the potential shown in Figure 5.9*

high volume fraction. This closely corresponds to the hard sphere order–disorder transition. It is likely that microstructural changes are responsible for this. In the region where the square law was obeyed there was a significant difference between the energy obtained from the slope, which was $210k_BT$, and that for the system. The origin of the difference between the model and the experiment is unlikely to be due to the collision frequency factor alone (assumed to be 1, experiment and theory equate when $v = 0.036$). The difference owes more to the simplicity of the model. The model has little information on the microstructure in the system simply because it is assumed that as soon as it is formed it is destroyed. The inclusion of the structural relaxation processes is an important issue in improving the model. Nonetheless it preserves the concentration dependence and provides a good qualitative description of the system. Systematic variation of the interaction energies rather than the particle concentration has also been investigated. This can be achieved by taking a latex and varying the electrolyte concentration to alter the depth of the secondary minima. It is also possible to achieve this using a depletion potential. Here non-adsorbing polymer is added to a dispersion. The particles diffuse and on occasion the separation becomes less than twice the radius of gyration of the coil. This leads to attraction between the particles. When a critical concentration c^* is exceeded the excess osmotic pressure due to the polymer induces an attractive energy and above a critical particle concentration a yield stress. The yield stress is proportional to the osmotic pressure of the polymer Π and the concentration of free polymer c. If a linear relationship exists between the osmotic pressure and the concentration of polymer then a linear relationship will also exist between yield stress and polymer concentration:

$$\sigma_B \propto \Pi \propto ck_BT \tag{6.53}$$

Such a relationship is exhibited when cis-polyisoprene, a non-adsorbing polymer, is added to PMMA/PHS particles in dodecane.[14] This is illustrated in Figure 6.12.

We can see here that as the polymer concentration is systematically increased so the yield stress increases accordingly. Experimental observations suggest the Bingham yield stress increases from a polymer concentration of about c^*. It should be emphasised that the added polymer shows no significant viscoelasticity in the frequency range examined, where the colloidal dispersion becomes strongly viscoelastic. This microstructural model has been developed using a Bingham plastic and for weak flocculation we would expect pseudoplastic behaviour. It is

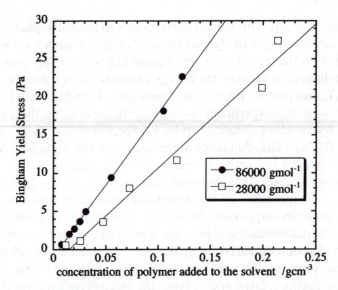

Figure 6.12 *Plot of the Bingham yield stress versus the concentration of cis-poly(iso-prene) added to a latex in dodecane (system HK4 with φ = 0.425, see Figure 6.6 for the codes)*

an extrapolated value in the same manner as can be achieved for hard spheres.

Perhaps the equation to receive most widespread use for the largest variety of systems is the Casson equation. This was introduced in Section 6.2 as a phenomenological model which is how it is often utilised. It has been widely used to describe flow curves for plastic and pseudoplastic materials over a limited range of stress and shear rates and can generate a far closer match to the curvature. The Casson approach is based on a phenomenological model. His argument is based on deforming the aggregate in a shear field where the particles begin to align in that field to form rod-like aggregates. Just as Michaels and Bolger ignored the Brownian contribution at high shear rates, Casson does the same, assuming the rate of making and the rate of breaking of rod-like aggregates can be attributed to two processes. The formation of the aggregate is due to the interparticle attractive forces and its disruption due to the hydrodynamic forces. As an ever-increasing shear field is applied the mean axial ratio falls until in the limit it approaches that of the primary particle forming the aggregate. Suppose we consider the case of spherical particles forming the primary species then we obtain for the viscosity and the yield stress:

$$\eta_{pl} = \eta_0(1 - \varphi)^{-u} \qquad (6.54)$$

$$\sigma_{\mathrm{C}} = \left[\left(\frac{\eta_{\mathrm{pl}}}{\eta_0}\right)^{1/2} - 1\right]^2 \frac{k_{\mathrm{B}}T}{b_1 a^3} D\bar{F} \qquad \text{with } D = \left[-\frac{\ln(1-\varphi)}{\ln(\eta_{\mathrm{pl}}/\eta_0)}\right]^2 \qquad (6.55)$$

where u is a constant controlled by the axial ratio of the rod and b_1 is a constant approximately equal to 24. There is an interesting similarity in the form of the Casson yield stress and the extrapolated Krieger yield stress (Equation 6.7) indicating in part the reason for the utility of the Casson expression. Interestingly, scattering and computer simulation which post-date the model by many years support the idea of 'string' formation in some systems, something akin to the formation of a rod. The data in Figure 6.13 shows the Casson equation (6.3) being used to linearise the flow curve of the system shown in Figure 6.11. This is at a volume fraction of 0.36.

The extrapolated yield stress gives 0.06 Pa and a plastic viscosity of 3.88 mPa s. We can use this to estimate the force between the particles, which gives $425k_{\mathrm{B}}T/a$, in fair agreement with the value determined using pair potential curves. Here the Casson model has been used to partially linearise a pseudoplastic system rather than a system with a true yield stress.

Both the Casson model and that of Michaels and Bolger form a class

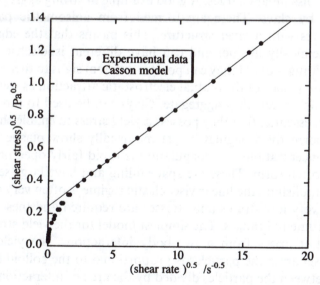

Figure 6.13 *A flow curve partially linearised using the Casson model for a weakly attractive system with the pair potential shown in Figure 5.9. The curvature at low stresses is indicative of a viscoelastic liquid. The Casson model successfully linearised the data where the particles can be visualised as aligning with the applied shear field. This would suggest almost complete breakdown of the aggregates*

of solutions for describing shear thinning. These two models are particularly attractive in that they relate to the colloid interaction forces, which can be ascertained through careful characterisation. They also include rate constants controlling the breakdown and formation of aggregates in a shear field. Establishing these terms provides the greatest difficulty. Whilst these models lack the rigorous detail embodied in linear viscoelastic theory they preserve the notion of structural breakdown and reformation in the shear field. The temporal changes in the system become increasingly important as the systems become strongly aggregated. The distortion of the floc structure plays an important role in energy dissipation mechanisms. The inherent thixotropy and structural irreversibility as interaction strengths increase means that strongly aggregated systems require a different description of their behaviour.[15]

6.3.5 Strongly Aggregated and Coagulated Systems

Colloidal particles are essentially attractive, so without the presence of a repulsive barrier to prevent their close approach they will coagulate. The interaction between the particles is typically dominated by the van der Waals' interactions, although strong aggregation can be induced by the attraction of dissimilar charges. A good example of strong aggregation is that shown by clays. These are formed from anisometric particles, typically plates with a layer structure. This means that the edges and faces are chemically distinct and can have different isoelectric points. Over certain ranges of pH they can possess different signs to their charges and as a consequence of the mutual electrostatic attraction as well as the van der Waals' forces, they aggregate. Clays can be used to throw pots and it is then essential that they possess a yield stress to enable shapes to be formed correctly. Coagulated systems usually show plastic flow at moderate concentrations. As coagulation is rapid fairly open networks of particles often form. These are space filling and have large sediment volumes. Establishing the linear viscoelastic regime is often very difficult as typically very low strains and stresses are required and these can be outside instrumental ranges. The simplest model for the yield stress is to assume that it arises from a two-body interaction. The yield stress required to separate the particles is proportional to the colloid interaction force between the particles divided by the area of interaction:

$$\sigma_y \propto \frac{F}{a^2} \varphi^2 \tag{6.56}$$

Russel *et al.*[16] pointed out that this approach is over-simplistic in not allowing for microstructural changes in the network as the concentration increases. However, the nature of the aggregate structure is path dependent, so the way in which the system is prepared is often critical in determining its microstructural nature and hence flow profile. A study of coagulation under shearing fields performed by Mills *et al.*[17,18] was coupled with freeze fracture electron microscopy measurements. A volume fraction range of $0.05 < \varphi < 0.25$ was used to produce coagu-lated latices with starting particle diameters in the range 250 nm $< 2a <$ 1250 nm. Under constant stress it was shown that these produce spherical aggregates with a diameter of 8–10 μm regardless of the initial volume fraction. However, it was sensitive to the initial particle size increasing from 5 μm to 15μm for an increase in primary particle size from 250 nm to 1250 nm. They display a form of time dependence with the viscosity dropping from a maximum value with time to a constant viscosity. This is seen even when systems are taken above their gel point. The drop in viscosity appeared permanent with aggregates having a packing fraction of $\varphi_{mf} = 0.55$. The greatest viscosity drop is seen with the highest concentration systems. These systems become pseudoplastic rather than plastic after shear conditioning. The notion that aggregates with a spherical flow volume occur under shear can be coupled with the Michaels and Bolger model used for weakly attractive systems. The idea is straightforward. Suppose we apply a shear field. In a strongly aggregated system the hydrodynamic forces are not great enough to break the network down to its individual particles. This is easily recognised because the plastic viscosity of such a system will be much greater than that of a system of well dispersed particles at the same concentration. The applied stress breaks up elements of the network and these can be thought of as forming spherical flow units with a volume fraction φ_f. The high shear viscosity is thus given by the Dougherty–Krieger expression:

$$\eta = \eta_0 \left(1 - \frac{\varphi_f}{0.62}\right)^{-[\eta]0.62} \tag{6.57}$$

We have assumed that a random packed structure is more likely at high rates with a distribution of floc sizes. The volume fraction of flocs will depend upon the floc packing fraction, giving rise to a floc diameter $2a_f$ with n_f particles per floc:

$$\varphi_f = \frac{\varphi}{\varphi_{mf}} \quad \text{and} \quad n_f = \varphi_{mf}\left(\frac{a_{eff}}{a}\right)^3 \tag{6.58}$$

The Bingham yield stress is thus

$$\sigma_B = \frac{24v|V_m|\varphi_f^2}{\pi^2(2a_f)^3} \tag{6.59}$$

This model provides us with a useful way of describing the flow behaviour of a strongly aggregated system which breaks down into spherical aggregates. The problem with this approach is that it is not readily predictive of material behaviour or even a completely accurate representation of all the structural changes or energy dissipation mechanisms possible.[15] However we can use experimental data to obtain values for the floc volume as the interaction forces or microstructure are changed. A good example is kaolin which consists of clay platelets. A typical flow curve for a kaolin suspension at pH 4 is shown in Figure 6.14.

As the pH is adjusted or surfactants are added the kaolin plates can aggregate as edge-edge, edge–face or face–face in quiescent conditions (Figure 6.14). Edge–face aggregation gives rise to a 'house of cards' structure. For each system both the concentration and the type of aggregation can be modified. In Figure 6.14 two flow curves can be seen. The system with the higher yield stress is aggregated in the 'house of cards' structure and shows a significant yield stress. However, when the surfactant hexadecyl trimethyl ammonium bromide (HDTAB) is added

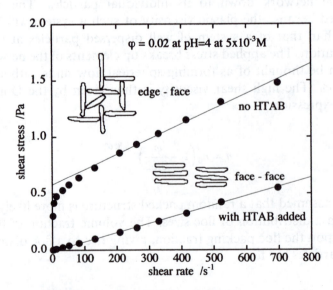

Figure 6.14 *Flow curves for sodium kaolinite at pH 4. The two systems are differ only by the addition of hexadecyl trimethyl ammonium bromide (HDTAB)*

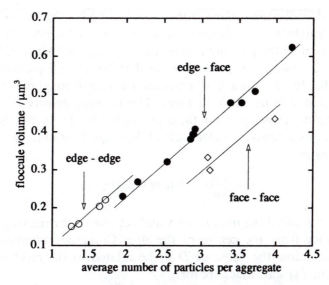

Figure 6.15 *The floc volume in shear against the number of particles per floc for a kaolin suspension. Different aggregation numbers for edge–edge and edge–face interactions were achieved by adjusting pH and electrolyte conditions. Face–face interactions were induced by the addition of a monolayer of HDTAB*

it adsorbs on the plate faces, disrupting the structure. This can be seen as a reduction in the yield stress. This changes the size of the floc in shear. The relationship between the calculated volume of the aggregate in flow versus the number of particles in the aggregate is shown in Figure 6.15.[19]

In shear the edge–edge and edge–face aggregation is very similar in behaviour in shear, suggesting that there is a degree of shear alignment in the flocs. Face–face aggregation is distinctly different, showing a lower volume for the same number of particles. Intuitively this is in agreement with expectations in that edge–face aggregation would be expected to have the larger volume per number of particles than edge–face. So whilst it is true these models lack the rigour of those constructed from linear viscoelastic approaches they do represent a self-consistent and useful picture.

There are many other forms of clay including synthetic clays such as laponite, a synthetic hectorite. Suitable tuning of the properties of these systems can produce similar structures. These systems have the advantage of small particle size and a relatively improved level of particle monodispersity over their naturally occurring counterparts and are being increasingly used as rheology modifiers.

The aggregation of spherical particles such as small particles of silica have been modelled by a fractal approach.[16,17] Fractal mathematics can be used to describe the surface or the mass of a body. The idea of a mass fractal is straightforward. Imagine you have an isolated floc of radius a_f where we assume the packing is constant throughout that floc. We can determine the number of particles n_f in that floc as in Equation (6.58). Suppose the floc does not have constant packing throughout its structure, but that it is dendritic in form. The packing density of the floc begins to reduce as you go from the centre of the floc to its edge. Suppose this reduces as a constant power law D. We can determine the number per floc now as

$$n_f = \left(\frac{a_f}{a}\right)^D \quad \text{where } 0 < D \leq 3 \tag{6.60}$$

The term D is called the *fractal index* and represents the packing change with distance from the centre of the floc. Computer simulation and experiments allow the value of D to be related to the mechanism of aggregation. Typical values are for:

1. Rapid aggregation, called *Diffusion Limited Aggregation* (DLA), when species touch they stick. Particle–particle aggregation gives $D \approx 2.5$; aggregate–aggregate aggregation gives $D \approx 1.8$.
2. Slow aggregation, called *Rate Limited Aggregation* (RLA), species have a lower sticking probability. Some are able to rearrange and densify the floc, $D \approx 2.0$–2.2.

The lower the value of D the more open the packing. Suppose we now imagine a dispersion which has been driven to instability by, say, changing temperature. We can visualise sites for the nucleation of flocs occurring randomly throughout the whole volume of the dispersion. The total number of primary particles remains unchanged so we can determine the volume fraction of flocs:

$$\varphi_f = \varphi \left(\frac{a_f}{a}\right)^{3-D} \tag{6.61}$$

Given that there is a relationship between the yield stress and the concentration we might expect a power relationship between them reflecting the fractal index, *i.e.* $\sigma_y \propto \varphi^m$.

There is a concentration limit to the applicability of the fractal approach. As simultaneous nucleation occurs the flocs will grow and as they grow they may eventually touch. Further growth would require the interpenetration of the flocs and when this occurs the idea of a fractal floc is lost because the structure changes. We can roughly estimate the limits on this by assuming the flocs reach a packing fraction before interpenetration. We need at least three particles in an aggregate otherwise connectivity is not possible. The number in the nucleus of the aggregate we can approximate as $(a_f/a) = 3$. So for a floc packing of say 62%, the maximum volume fraction without interpenetration is given by

$$0.62 = 3^{3-D}\varphi \tag{6.62}$$

So it is possible to determine a maximum limit where the fractal structure begins to interpenetrate. For the various values of D we find

$$D = 1.8 \quad \varphi_{max} = 0.17$$
$$D = 2.0 \quad \varphi_{max} = 0.21$$
$$D = 2.5 \quad \varphi_{max} = 0.36$$

Once interpenetration occurs the resistance to deformation increases markedly, so for example we would expect compaction of a sediment to become limited, as would further concentration in a filter press. It is worth emphasising the point that this is a simplistic approach, as prior to interpenetration the clusters undergo structural rearrangements changing their fractal index at a critical volume fraction. A typical data set for yield stress is shown in Figure 6.16.[19]

Figure 6.16 shows a clear transition between the low volume fraction response and that at higher volume fractions. Perhaps the major problem with a fractal model is that it underestimates the complexity of colloidal

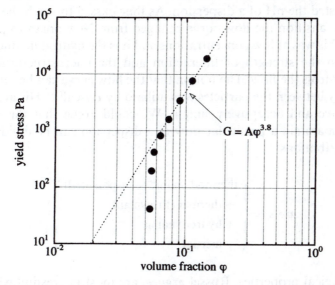

Figure 6.16 *Plot of yield stress versus concentration for a coagulated latex, perhaps indicating a transition between low and high volume fraction behaviour*

microstructures in many aggregated systems. This was well described by Dickinson[20] when he considered the order relative to the length scales. Over short ranges, say one or two particle diameters, the structure is akin to a molecular liquid in order. At volume fractions around the gel point, say $\varphi_g = 0.05$, for between two and six particle diameters fractal order is observed. At larger distances the particle density has reduced to the average for the dispersion. At high volume fractions ($\varphi > 0.25$) a more liquid-like structure is maintained and no fractal order is observed. This is an area ripe for further work but the difficulty and path dependence of the microstructure formed presents awesome difficulties in obtaining a generalised picture of the behaviour.

6.3.6 Long-range Repulsive Systems

The colloid stability map (Figure 6.9) shows a stable region where charge repulsion dominates over all other pair interactions between the particles. This arises at low levels of electrolyte and/or high levels of surface charge. The mutual repulsion between the particles leads to microstructural changes as a function of concentration. A liquid-like order of colloidal particles is replaced by a 'crystalline order' with regular interparticle spacings. These changes occur at relatively low concentrations. The non-linear responses of charge-repulsion systems is related to the initial microstructure. This was demonstrated by Lindsay and Chaikin[21] who measured the shear stress as a function of shear rate as they adjusted the pH of a dispersion. As they passed through the order–disorder transition the flow curve changed from Newtonian to pseudoplastic. This was at low concentrations and it is the hydrodynamic forces relative to the strength of interaction and the microstructural order which controls the flow. The notion that the flow curve can be related to the energy between the particles was utilised by Buscall.[22] His argument is attractive and deceptively simple. We could argue that the applied stress to maintain a particular flow in a dispersion can be divided into four contributions:

$$
\text{stress} = \left\{
\begin{array}{l}
\text{thermodynamic (pair potential)} \\
+ \text{ thermal (Brownian)} \\
+ \text{ hydrodynamic} \\
+ \text{ inertial}
\end{array}
\right\}
$$

Rheological properties, Russel argues, are most interesting when the pair interaction potential between the particles dominates the flow properties, and this is the case Buscall is interested in investigating. Providing we are working in well defined flows at low Reynolds numbers we can ignore inertial effects. Suppose we have a particle with a hard core and a repulsive potential $V(R)$ decaying away from that core. The 'soft tail' of this potential opposes the close approach of particles in a shear field in comparison with a hard sphere. Thus from the viscosity we can define an effective volume fraction φ_{eff}

$$\frac{\eta(\sigma)}{\eta_0} \cong \left(1 - \frac{\varphi_{\text{eff}}(\sigma)}{\varphi_{\text{m}}}\right)^{-[\eta]\varphi_{\text{m}}} \tag{6.63}$$

This volume fraction represents an increase in the effective particle size to give a collision diameter R_{eff}

$$\varphi_{\text{eff}}(\sigma) = \left(\frac{R_{\text{eff}}(\sigma)}{2a}\right)^3 \tag{6.64}$$

This new particle diameter represents the separation at which the hydrodynamic stress balances with the thermal and thermodynamic stresses:

$$\text{thermodynamic} = \text{hydrodynamic} + \text{thermal}$$

Buscall suggests the following prescription:

$$V[R_{\text{eff}}(\sigma)] = \sigma_r \left(\frac{\varphi_{\text{eff}}(\sigma)}{\varphi}\right) h[\varphi_{\text{eff}}(\sigma)] k_{\text{B}} T + \frac{1}{2} k_{\text{B}} T \tag{6.65}$$

where $h[\varphi_{\text{eff}}(\sigma)] = [0.032 + 1.04\varphi_{\text{eff}}(\sigma)]^{-1}$ is the reciprocal of a dimensionless characteristic stress found by fitting quasi-hard sphere experimental data. So a measurement of viscosity at different stresses can be used to calculate the interparticle potential as a function of separation. This approach has the great strength that it is able to reduce data from many experiments by referencing them to experimentally measured systems. It confirms to an extent that pseudoplastic curves belong to the same family of behaviour. The model could be developed by taking account of the structural changes under shear with the packing fraction being dependent on applied stress. Incorporating structural changes into models of strongly repulsive systems is difficult. Zukoski and Chen[23] incorporated the specific form of the colloid pair potential in order to predict a critical strain at which non-linear responses will be observed, and they found $\gamma_{\text{crit}} \approx \mathbf{O}(0.04)$.

The form of the non-linear viscoelastic curve for storage and loss modulus shown in Figure 6.5 has also been investigated. The structure of the strongly repulsive systems is typically fcc, being formed from many ordered domains. We can imagine that as each domain is strained the boundary where one grain meets another begins to slip. Whilst this oscillating strain is maintained there is a disordered region which is liquid-like. As the amplitude of the strain increases the volume fraction of liquid-like regions increases. We could argue that the compliance of the total sample is formed from the sum of liquid and solid-like zones:[9]

$$J'(\gamma) = J'_S \varphi_S(\gamma) + J'_L \varphi_L(\gamma) \tag{6.66a}$$

$$J''(\gamma) = J''_S \varphi_S(\gamma) + J''_L \varphi_L(\gamma) \tag{6.66b}$$

where L represents the liquid-like and S the solid-like properties. In the low strain limit we have the values for the solid:

$$J'(\gamma \to 0) = J'_S \quad \text{and} \quad J''(\gamma \to 0) = J''_S \tag{6.67}$$

We can think of the strain as inducing melting. At the melting temperature T_M we would expect the volume fraction of each phase to be equal to 0.5. We could argue that this happens where the storage and loss moduli are equal. Given these assumptions we can calculate the amount of solid and liquid-like material present as a function of strain. The apparent volume fraction of liquid is shown for a polyvinylidene fluoride latex in Figure 6.17.

The data superimposes onto the same curve when plotted as the applied strain over the melting strain. We can think of this as the temperature relative to the melting temperature. The model is simple and the agreement between data from different systems is also excellent. This is indicative of the similarity in the form of the storage and loss moduli for these systems. As the strain is increased the response becomes dominated by flow, typically above $1.7\gamma_m$. The crudity of this model reflects the limited attempts that have been made to model the non-linear viscoelasticity of dispersions. Stress relaxation studies at finite strains are also rather limited. Figure 6.18 shows the behaviour of a concentrated silica dispersion as the strain is increased. There is a suggestion that at long times a low frequency modulus or $G(0)$ value is present. This sample does behave as a linear viscoelastic solid although the magnitude of $G(0)$ could not be determined in the instrumental measurement range.

As we have seen in the previous chapter, strongly interacting colloidal systems are characterised by a very rapid increase in the low shear

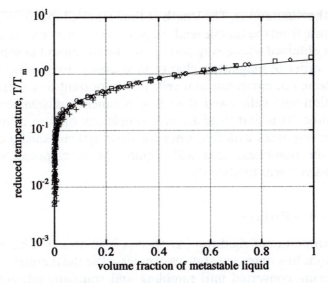

Figure 6.17 *Plot of the pseudo-melting curve calculated from a strain sweep. This data is for a polyvinylidene fluoride latex. Similar curves can be obtained for silica and polystyrene latex dispersions*

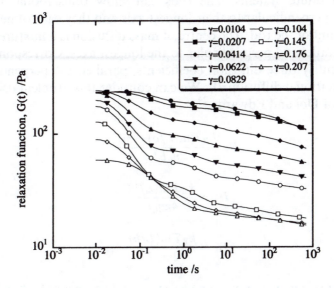

Figure 6.18 *The stress relaxation at a range of strains for a silica dispersion*

viscosity with increasing concentration. They display shear thinning but the high shear rate plateau is often difficult to reach. Large particles tend to show such a rapid change with concentration between the viscoelastic liquid and viscoelastic solid behaviour that flow curves are only obtained for a very limited range of concentrations. The concentration sensitivity imposes other limitations. The length of time required to obtain a steady state response from the sample tends to place restrictions on the accuracy of the data obtained where evaporation of a small amount of solvent can lead to an order of magnitude change in viscosity. Even when samples are constrained in environmental cells it is important to determine the concentration both before and after the experiment has been performed. Another difficulty is that there are only simple descriptions of the linear viscoelastic responses with frequency for these systems and so developments in the non-linear area will require the development of more comprehensive linear treatments.

6.3.7   Rod-like Particles

The flow properties of rigid rod-like species is an area that has received less attention. In a quiescent state we can visualise the thermal energy of the rods being converted into tumbling and translational motion. In order to induce flow and a change in viscosity an applied shear field must overcome this Brownian movement. We can identify three types of motion in dilute systems. The rods can show translational motion parallel or perpendicular to their longest axis and they can rotate about the mid-point of the rod. The centre of mass diffusion is a mixture of the motion parallel and perpendicular to the longest axes. Correspondingly we can identify three diffusion coefficients, parallel, $D_{\parallel}$, perpendicular, $D_{\perp}$ and rotational diffusion D_r. We can consider this problem using the approach of Doi and Edwards.[24]

$$D_r \frac{3k_B T[\ln(L/b) - c_1]}{\pi \eta_0 L^3} \tag{6.68}$$

$$D_{\perp} = \frac{k_B T \ln(L/b)}{4\pi \eta_0 L} \tag{6.69}$$

$$D_{\parallel} = \frac{k_B T \ln(L/b)}{2\pi \eta_0 L} \tag{6.70}$$

for a rod of length L and diameter b. The term $c_1 \approx 0.8$ for a simple rod, but its value depends upon the details of the rod geometry. As we increase the concentration the motion of the rods will become restricted as they interact with each other. We can visualise four concentration regimes based on the number density of rods ρ.

1. Dilute solution $\{\rho_1\}$. The concentration of the rods is such that the

average spacing between them is much larger than the longest dimension of the rod:

$$0 \leq \rho_1 \leq L^{-3}$$

2. Semi-dilute solution $\{\rho_2\}$. As the concentration rises the rods will begin to interact and their diffusion will become restricted. However, we have not allowed for the excluded volume of the rod and have treated it as a line with no thickness. The excluded volume is of the order of bL^2 and until the concentration of rods is such that the particles overlap into this excluded volume the spatial distribution of rods is relatively unaffected:

$$\rho_1 \leq \rho_2 \leq (bL^2)^{-1}$$

3. Concentrated (isotropic) solution $\{\rho_3\}$. Once the concentration exceeds ρ_2 the polymers have a tendency to align with each other. They still show a degree of disorder but over a limited concentration range both the dynamics and distribution of particles in space are substantially modified. This isotropic state is maintained up to a concentration ρ^r which is of the same order as $(bL^2)^{-1}$:

$$\rho_2 \leq \rho_3 \leq \rho^r$$

4. Liquid crystalline solution. A phase transition occurs at this point and the system shows a high degree of order. These systems inevitably have significantly different features from those above.

As with spherical particles the Péclet number is of great importance in describing the transitions in rheological behaviour. In order for the applied flow field to overcome the diffusive motion and shear thinning to be observed a Péclet number exceeding unity is required. However, we can define both rotational and translational Péclet numbers, depending upon which of the diffusive modes we consider most important to the flow we initiate. The most rapid diffusion is the rotational component and it is this that must be overcome in order to initiate flow. We can define this in terms of a diffusive timescale relative to the applied shear rate. The characteristic Maxwell time for rotary diffusion is

$$\tau_r = \frac{1}{6D_r} \tag{6.71}$$

So our rotary Péclet number is given by

$$Pe_r = \dot{\gamma}\tau_r = \frac{\dot{\gamma}}{6D_r} \tag{6.72}$$

Once we reach the semi-dilute regime and the diffusive motion becomes restricted the Péclet number becomes concentration dependent. The Kerr effect has been used to demonstrate this. When an electric field is applied to a system of uncharged rods they tend to align with that field due to the anisotropy of their dielectric properties. If the field is released the rods diffuse back to their random configuration. If laser light is allowed to pass through the sample, the rate at which the rods diffuse back to their equilibrium concentration can be monitored by the exponential decay of intensity with time after the field is removed. It is dependent upon their length relative to the concentration. This has been shown by both computer simulation and experiment. In the semi-dilute regime modelling the restriction for the diffusive motion as the concentration increases can be achieved by constraining the rod-like molecule to a tube. The dimensions of the tube are controlled by the concentration of the particles and the length of the rod. For a tube of radius a we have

$$aL^2 \approx \frac{1}{\rho} \tag{6.73}$$

The rod is visualised as being constrained to a tube in a similar fashion to entanglements constraining a polymer in reptation theory. So for a finite concentration our diffusion coefficient and rotary Péclet number changes:

$$D_{rc} \approx D_r\left(\frac{a}{L}\right)^2 \text{ or as an equality } D_{rc} = \beta D_r\left(\rho L^3\right)^{-2} \tag{6.74}$$

where β is a numerical factor typically of order 10^3. Comparing models for hard rods with no pair interactions with 'real' colloidal systems introduces some complex behaviour. One approach to test the concentration dependence of the rotary diffusion coefficient would be to use the viscosity of the dispersion to replace the solvent viscosity in the expression for rotary diffusion. Rod-like species, particularly in an aqueous environment, will inevitably have a complex distribution of charges and the specific nature of the pair interactions of the rods are important. This is illustrated in Figure 6.19 where the Kerr effect has been used to determine τ_r in infinite dilution. It is plotted as a function of the viscosity of the solvent.

Here PTFE particles with an axial ratio of about 2.4 were dispersed in different concentrations of glycerol in water. We might predict that as the

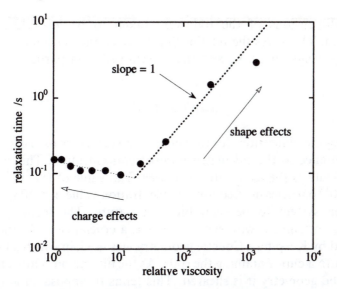

Figure 6.19 *Kerr effect on PTFE particles. The Kerr relaxation time is plotted against the viscosity of a water/glycerol mixture relative to the viscosity of water (L = 388 nm and b = 167 nm)*

viscosity of the solvent is increased the relaxation time would follow the line with a power law slope of 1. The reason that it does not follow this relationship at low concentrations is because the particles are charged. As the viscosity is changed by the addition of glycerol, the dielectric constant of the medium changes, and the charge distribution around the particles also changes, influencing the relaxation of the particles relative to their ionic atmosphere. Most anisotropic colloidal systems will be far from ideal rigid rods. This does not negate the tube model. Experiments on rigid polymers do provide insights. The Carreau[25] equation for example has been used by Chauveteau[26] to describe the viscosity of a polysaccharide:

$$\eta(\dot{\gamma}) = \frac{\eta(0)}{\left[1 + (\tau_r\dot{\gamma})^2\right]^m} \qquad (6.75)$$

Here τ_r is a characteristic relaxation time. The effectiveness of this fitting with a characteristic relaxation time is encouraging. It suggests that the flow properties are most appropriately described by a characteristic relaxation time rather than a critical stress. The behaviour of high concentration systems of particles has been studied by Keeping[27] using a natural clay, attapulgite, which occurs as rods with high axial ratios. There is an inevitable degree of polydispersity with natural systems. The average dimensions of the rod were taken to be $b = 0.13 \, \mu m$ and $L = 2.8 \, \mu m$. This gives the relationship between the percentage concentration by weight used, w, and the concentration in particles per unit volume:

$$\rho = 0.1 \times 10^{18} w \qquad (6.76)$$

Defining the dimensions for the rod and the geometry used can make a large difference to the calculated number concentration. This becomes more difficult as the axial ratio increases. The value of $(bL^2)^{-1}$ is very close to 10^{18} particles m^{-3} so for a concentration value $w < 10$ we would expect the system to be semi-dilute in nature. The loading of the measuring instrument with such systems is a crucial issue. In the study performed by Keeping a Couette geometry was used, in which a bob was lowered into a cup containing the rods. As the dispersion is forced up the sides of the geometry it is sheared. This tends to impose an alignment upon the rods and a strain to be stored in the sample which is relaxed very slowly. A short period of applied shear results in the loss of alignment and the loss of most of the residual stress retained in the structure. When the viscosity was measured it was done using a controlled stress instrument and this led to a rapid change in viscosity with stress. The viscosity at any stress was found to increase with concentration and to give comparatively large values compared to an equivalent concentration of spherical particles. The concentration dependence of the low shear viscosity of these systems is very difficult to obtain with high precision. The relaxation times tend to be long. The concentration dependence of the viscosity in the semi-dilute regime is given by[24]

$$\eta(0) = \frac{\rho k_B T}{10 D_{\text{rc}}} \qquad (6.77)$$

For a system with all the parameters fixed except concentration we would expect the relationship:

$$\eta(0) = K w^3 \qquad (6.78)$$

to hold true. From the findings of Chauveteau, if we assume that $\tau_r \dot{\gamma}$ is much larger than unity then we might expect the following relationship to hold:

$$\frac{\eta(\dot{\gamma})}{K w^3} = (\tau_r \dot{\gamma})^{-2m} \qquad (6.79)$$

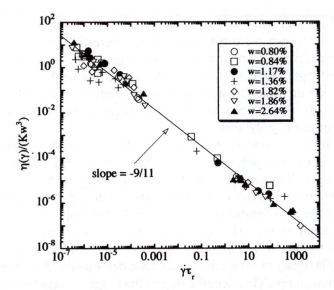

Figure 6.20 *Plot of the normalised viscosity against the normalised shear rate for attapulgite rods. The viscosity is normalised to allow for concentration and the shear rate is normalised by the relaxation time*

This reduces the data onto a master curve at moderate to high shear rates and this is shown in Figure 6.20 for a range of concentrations.

The value of K has been arbitrarily selected to normalise the data and will include colloid interaction terms and polydispersity. So we expect that in the semi-dilute regime the viscosity of interacting rods is reasonably well described by the tube model. The relaxation times required to normalise the data increase with concentration roughly as w^2 as expected, although the scatter on the data is significant. A curious feature of the data for rods emerges when we examine the rate of shear thinning. The power law of the slope is given by $2m = 9/11$. This initially seems a surprising finding because this is in agreement with the predictions of Graessley[28] who obtained this limiting slope for a model for the entanglement of high molecular weight polymers. This tends to confirm the observations of Doi and Edwards[24] who suggest that the shear thinning properties of rods are similar in form to highly entangled polymers.

6.4 THE MICROSTRUCTURAL APPROACH – POLYMERS

There are a wealth of experiments, papers and texts covering the non-linear viscoelastic response of polymers. In this book we cannot hope to cover the area in as much detail as the many admirable texts in this field.

Just as Russel *et al.*[16] has argued for particles that it is the pair interactions between them that make them interesting, so we can argue that polymers really start to become interesting once they interact and start to entangle. Thus the semi-dilute and concentrated regimes are where the polymer scientist will find most excitement. We will begin with simple ideas for homopolymers and then begin to consider the role of crosslinking and copolymers.

Models for dilute systems have been developed from the Rouse model for linear viscoelastic polymers. The linear viscoelasticity is described by a sum of Maxwell models. The elasticity of the spring in these models can be made strain-dependent, resulting in a shear rate dependence for dilute systems. Models for flexible elastic dumb-bells deforming in a strain field produce similar results. However, once the polymers begin to interact and coupling between chains commences, a different approach is required. The onset of interactions has been demonstrated by Freed and Muthukumar to result in a lengthening of the longest relaxation time and hence an increase in viscosity. The extent of this effect depends on the expansion of the chain in the solvent. As the concentration of chains is increased further the chains will begin to come into contact and start to interpenetrate. At this point chains have the opportunity to become entangled. This gives rise to an increase in the viscoelastic properties.

6.4.1   The Role of Entanglements in Non-linear Viscoelasticity

The exact nature of an entanglement has been the subject of extensive debate. We should clearly distinguish between this type of interaction and, say, a transient network. A network for example might arise because of the affinity of blocks of chemical groups on neighbouring chains. Such an association is driven by the interaction energetics, and is enthalpic in nature. When considering entanglements we must primarily concern ourselves with the topology of the polymer and understand that constraints in the motion of the chain arise due to one chain looping itself around elements of another. The simplest case to consider is that of a polymer with no solvent present at a temperature above its melting point, where it is free to entangle. A model developed by Graessley[28] is the simplest approach to give a realistic microstructural description of the process and return an expression that reproduces experimental data. Consider the polymer in the low shear rate limit. The polymer is able to diffuse as if in its quiescent state so the formation and loss of entanglements is unperturbed. However, as the shear rate is increased to a critical value the polymer begins to display shear thinning. The reciprocal of the characteristic rate at which this occurs defines a characteristic time τ_η.

Graessley has argued that once this characteristic rate is exceeded, a new steady state response is achieved where the number of entanglements between polymer chains has reduced. This is because, as the chains move past each other, the characteristic time for the formation of the same number of entanglements in the quiescent state requires a longer time than the shear field allows the chains to remain in contact. The time required to form an entanglement and the energy dissipated is proportional to the viscosity at that rate. Graessley defines two key terms which control the viscosity. The first term h_1, the basis of an earlier model, is the rate of energy dissipation of a chain of fixed length and frictional drag relative to those values in the zero shear rate limit:

$$h_1 = \frac{2}{\pi}\left(\text{arc cot}\,\theta + \frac{\theta(1-\theta^2)}{(1+\theta^2)^2}\right) \qquad \text{where } \theta = \frac{1}{2}\tau_e\dot{\gamma} = \frac{\eta(\dot{\gamma})}{\eta(0)}\frac{1}{2}\tau_\eta\dot{\gamma} \quad (6.80)$$

The time τ_e represents the entanglement formation time at the current shear rate. The second term g_1 is the average number of entanglement sites for a chain of fixed length in steady flow relative to the number in the zero shear rate limit:

$$g_1 = \frac{2}{\pi}\left(\text{arc cot}\,\theta + \frac{\theta}{(1+\theta^2)}\right) \qquad (6.81)$$

The viscosity for a monodisperse polymer can be shown to be given by

$$\frac{\eta(\dot{\gamma})}{\eta(0)} = h_1 g_1^{3/2} \qquad (6.82)$$

This deceptively simple expression is capable of describing the shear thinning response of monodisperse polymers with a high level of precision. In the high shear limit we obtain

$$\eta(\dot{\gamma}) \propto \dot{\gamma}^{-9/11} \qquad (6.83)$$

a power law which has been observed for many systems.[29] The Graessley prediction can be seen in Figure 6.21.

A numerical implementation of this approach can be generalised to include the polydispersity of the polymer. As polydispersity is increased the power law of $-0.8181\ldots$ reduces. The onset of shear thinning, where $\tau_\eta\dot{\gamma}^{-1} = 1$, results in a slightly lower viscosity for polydisperse systems at this rate. So far we have neglected the origin of the characteristic time τ_η for the system which we would like to describe in terms of the chemical

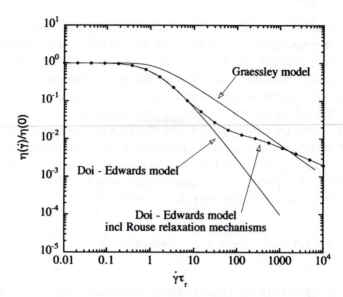

Figure 6.21   *Comparison of Graessley and Doi–Edwards models for normalised viscosity versus normalised shear rate. Also shown is an estimate of the role of short time Rouse relaxation mechanisms within the tube*

properties of our polymer. This is difficult to establish because there is a distribution of values. Graessley has calculated an average time for the relaxation of an entangled network, τ_{ave}. He assumes that the frictional drag of an entanglement site depends upon its location in the chain – the further from the chain ends the greater the drag. The associated elasticity also depends upon the position of the site along the chain relative to the ends. If the effective molecular weight of the chain between entanglements is denoted by M_e then he obtains[30]

$$\tau_{ave} = 1.8 \frac{\eta(0) M_e}{\rho_p RT} \tag{6.84}$$

where ρ_p is the mass density of the polymer. This term effectively represents the elasticity divided into the viscosity, *i.e.* a Maxwell relaxation time, multiplied by a constant representing an average from a relaxation spectrum. The value of M_e is where much of the specific chemical interactions of the polymer are contained and is characteristic of the polymer conformation and molecular architecture. Graessley predicts a 'narrow' relaxation spectrum for the entangled portion of the structure which might allow a single relaxation time to replace the spectral response. His average time is still a little too short to represent the onset of shear thinning and typically we observe experimentally[29]

$$3.5 \text{ to } 6 = \tau_\eta \frac{\rho_p RT}{\eta(0) M_e} \tag{6.85}$$

We would expect a difference between the average time and that which determines the onset of shear thinning since the low shear viscosity tends to be dominated by the longest mode of relaxation.

6.4.2 Entanglement of Solution Homopolymers

Suppose we now consider our highly entangled polymer and begin to introduce a good solvent to the network. If only a small addition of solvent is made there will be a reduction in the density of the system and we can imagine that the viscosity will change in proportion to the density, providing the same network structure is maintained. A region in which we have solvent insensitivity has been argued by de Gennes using scaling concepts to occur above a concentration c^t. This is a region where a near-homogeneous interpenetrating network is observed. Under these conditions we would expect the model of Graessley to still apply. However, as the mass of added solvent is increased and the network structure begins to change, the model is less appropriate. The nature of the change is going to depend upon the nature of the polymer–solvent interactions as compared with polymer–polymer interactions.

Models for concentrated solutions have been tackled by Williams and comparisons between experimental data and the Graessley approach have also been undertaken. Some general observations can be made for simple homopolymers in a good solvent. As the polymer concentration reduces, the zero shear rate viscosity falls and the shear rate at which shear thinning begins shifts to higher values so that τ_η reduces. Typically the region where the viscosity has a power law dependence on shear rate shows a value that progressively approaches zero as the concentration reduces, *i.e.* the value falls in magnitude from a plateau value of $-0.8181\ldots$ toward zero. A correlation between the chain overlap parameter and the value of the power law slope has been indicated by experiments. The chain overlap parameter is the product of the intrinsic viscosity and the concentration, $[\eta]c$, where the units are defined such that their product is dimensionless. As the concentration of the polymer is increased, $[\eta]c$ increases and the power law slope of the viscosity approaches $-0.8181\ldots$ Experimental observations suggest that

$[\eta]c > 30$ is required to reach the Graessley value so requiring a concentrated system.[32] The data roughly follows:

$$m = \frac{9}{11}\left\{1 - \exp\left(-\frac{[\eta]c}{15}\right)\right\} \qquad \text{where } \frac{\eta(\dot{\gamma} \to \infty)}{\eta(0)} \propto \dot{\gamma}^{-m} \qquad (6.86)$$

with the numerical terms in this expression provided from a curve fit to experimental data.

The chain overlap parameter has been used in combination with de Gennes' scaling argument to demarcate differing zones of behaviour, *i.e.* dilute, semi-dilute or concentrated (see Section 5.6). The value of $[\eta]c$ influences the extent of chain overlap, the potential to form entanglements, their density and as a consequence the nature of the non-linear viscoelasticity. We are repeating a very important point observed with particulates that good chemical and physical characterisation of the sample is a tremendous aid in identifying the phase behaviour and flow behaviour that might be expected. The relaxation time τ_η controlling the onset of shear thinning is also dependent upon the concentration region. At low concentrations of homopolymer in a good solvent it has been observed to behave as[29]

$$\tau_\eta \propto \frac{\eta(0)}{\varphi_p} M \qquad (6.87)$$

whereas with a high number of entanglements:

$$\tau_\eta \propto \frac{\eta(0)}{\varphi_p^2} \qquad (6.88)$$

where φ_p is the volume fraction of the polymer. The path of describing solution polymer rheology is littered with good intentions. It is particularly difficult with aqueous systems to develop general rules because this is where specific interactions tend to be the norm rather than the exception. The idea that we can universally use $c[\eta]$ to demarcate zones of behaviour is far from satisfactory, as has already been observed (Section 5.7). Currently the best approach open to us is to collapse the complex chemistry of the polymers into some well found parameters determined from experiment. These parameters, such as an effective tube dimension (reptation) or molecular weight between entanglements, should be defined in terms of a well found physical model. This approach will probably not allow us to be entirely predictive of the behaviour but it will provide insights into the physical processes that occur. If for example we know how changes in our reaction chemistry affect the molecular architecture then we can say something sensible about how these key parameters will change and thus how the polymer rheology will change.

6.4.3 The Reptation Approach

The scaling concept has proved one of the most powerful ideas in unifying different aspects of polymer physics. Whilst this approach enables us to determine the dependence of the rheological properties on molecular weight or concentration, for example, it does not allow us to explicitly determine flow profiles. The development of this approach for non-linear viscoelasticity was pioneered by Doi and Edwards[24] utilising reptation models for the polymer chain to describe the effects of a large strain. The starting point is the linear viscoelastic response of the chain which they assumed follows Rousian dynamics in a polymer melt. In an entangled structure we can visualise the polymer chain constrained to a tube. The tube follows a winding path through a network of surrounding polymer molecules. This tube has a contour length L which is shorter than the polymer, which will allow the polymer to fold back upon itself. The contour length follows the shortest path down the tube and can be thought of as a primitive chain composed of links. The regions where the real chain deviates from the path of the primitive chain can be thought of as defects that can move back and forth along the chain. These allow the tube to change with time and creep or reptate through the surrounding polymer. The tube has a dimension a which is controlled by the density of the network constraining the chain. It is represented in terms of the length of a link in the primitive chain. These parameters can be related to the basic dimensions of the real chain which has a link length b and number of links N:

$$La = Nb^2 \qquad (6.89)$$

The stress relaxation function $G_R(t)$ for a single chain is Rouse-like and given by

$$G_R(t) = \rho_c k_B T \sum_{p=1}^{\infty} \exp(-2tp^2/\tau_R) \qquad (6.90)$$

with the Rouse relaxation time, which is the longest mode in the relaxation spectrum of an isolated chain, being given by

$$\tau_R = \frac{\zeta N^2 b^2}{3\pi^2 k_B T} \tag{6.91}$$

We have expressed the relaxation behaviour in terms of the number of chains per unit volume. At this stage we are considering the polymer in an undiluted state. Suppose we now apply a step strain to the melt in the linear regime. Two different zones of behaviour can be seen relative to the time τ_e. This is the time at which the tube constraints begin to affect the relaxation of the chain:

$$\tau_e = \frac{\zeta a^4}{k_B T b^2} \tag{6.92}$$

The applied strain is affine and the whole of the tube is deformed along with the polymer. As the strain is infinitesimally small the contour length is unaltered. At very short times t after the strain is applied, $t < \tau_e \ll \tau_R$ the stress is relaxed as a Rouse chain. At short times we can make an approximation and replace our sum by an integral:

$$G_R(t) = \rho_c k_B T \int_0^\infty \exp\left(-2tp^2/\tau_R\right) \mathrm{d}p = \frac{\rho_c k_B T}{\sqrt{8}} \left(\frac{\tau_R}{t}\right)^{1/2} \tag{6.93}$$

At longer times when $t \geq \tau_e$ the constraints of the tube dimensions prevent further relaxation by Rousian rearrangements and reptation becomes the dominant process. In this time regime relaxation is characterised by two timescales: the relaxation of the contour length by τ_R and the tube disengagement time by τ_d. However, in the linear regime when $t > \tau_e$ the disengagement process dominates. The average length of the original contour is constrained to the middle portion of the tube and relaxation progresses as the ends of the tube wriggle free of their constraints. Progressively the average population of segments in the middle of the tube is reduced as more and more of the chain reptates free from the original tube. Within reptation theory we can define a function $\psi(t)$ that describes the probability of an element remaining in the tube. This is formed from a sum of only those odd-numbered relaxation modes that contribute to the reptation. If we multiple this by a network elasticity we find

$$G(t) = G_N \psi(t) = G_N \sum_{p;\mathrm{odd}}^{\infty} \frac{8}{p^2 \pi^2} \exp\left(-tp^2/\tau_d\right) \qquad \text{for } t = \tau_e \tag{6.94}$$

with the tube disengagement time

$$\tau_d = \frac{\zeta N^3 b^2}{\pi^2 k_B T}\left(\frac{b}{a}\right)^2 \tag{6.95}$$

Experimentally we know that the overlap between relaxation processes results in a smooth change in behaviour. This change in behaviour enables us to evaluate G_N. We know that when $t \approx \tau_e$ we change from Rouse $G_R(t)$ to reptation $G(t)$ behaviour:

$$G_N \cong G_R(t) \cong \rho_c k_B T\left(\frac{\tau_R}{\tau_e}\right)^{1/2} \tag{6.96}$$

Now all we need do is substitute our expressions for the relaxation times to give:

$$G_N \cong \rho_c k_B T\left(\frac{Nl^2}{a^2}\right) \tag{6.97}$$

We may use this to redefine our Rousian relaxation process:

$$G_R(t) \cong G_N\left(\frac{\tau_e}{t}\right)^{1/2} \qquad \text{for } t \leq \tau_e \tag{6.98}$$

These functions describe the linear response of the relaxation function, dividing it into two clearly defined although overlapping processes. Our primary aim in this section is to consider the action of large strains but before continuing along this path we need to consider how to relate this expression to experimentally measured parameters. The frictional term remains conceptually difficult for all but spherical particles (see Section 5.6). However the network modulus G_N has been studied for some years and can be related to M_e, the molecular weight between entanglement sites. In its simplest form, ignoring specifics of link functionality, it is the effective molecular mass that arises to describe G_N using a simple theory of rubber elasticity:

$$G_N = \rho_c k_B T\left(\frac{M}{M_e}\right) \tag{6.99}$$

So by equating the two network moduli we can obtain our tube parameters. If we include all the necessary numerical constants we get

$$\frac{M}{M_e} = \frac{4}{5}\left(\frac{Nb^2}{a^2}\right) \tag{6.100}$$

This relationship allows us to make a link with experimentally determined parameters such as entanglement density. This should apply in the melt state. In practice this relationship is not completely satisfactory and to an extent a must be regarded as an adjustable parameter. It should be recalled throughout this modelling that we have not included the nature of any specific interactions and in real systems these are likely to be determining factors. The low shear rate viscosity can be determined from the integral in the stress relaxation function. As it is dominated by the longest time process, it is quite satisfactory to calculate it from the reptation motion alone:

$$\eta(0) = \int_0^\infty G_N \psi(t) \mathrm{d}t = \frac{\pi^2}{12} G_N \tau_d \tag{6.101}$$

The tube disengagement time is the longest relaxation process. Now suppose we were to apply a large strain to the sample so that we are no longer in the linear regime. When the chain is deformed the contour length increases and the tube is deformed. At a time t equivalent to τ_R this part of the strain is relaxed and the contour length is that in the quiescent state. After time τ_d the primitive chain has disengaged itself from the surrounding tube by reptation. The deformation of the tube and the subsequent relaxation is a complex feature to model. In its simplest approach the Doi–Edwards solution results in a decoupling between the linear relaxation function and the strain dependence. We have seen such a decoupling approach before (Section 6.2.2):

$$G(t, \gamma) = G(t) \frac{f_1(\gamma)}{\gamma} \tag{6.102}$$

The function $f_1(\gamma)$ represents the non-linear strain dependence of the deformed tube. The non-linear stress relaxation function in the reptation zone is thus

$$G(t) = G_N \psi(t) \frac{f_1(\gamma)}{\gamma} = G_N \frac{f_1(\gamma)}{\gamma} \sum_{p;\text{odd}}^\infty \frac{8}{p^2 \pi^2} \exp\left(-tp^2/\tau_d\right) \tag{6.103}$$

This follows the form of phenomenological descriptions of polymer systems such as the BKZ model[10] which is encouraging. Very good fits to experiments have been found using this approach. In order to take this idea forward the most convenient method is to use the memory function:

$$m(t) = -G_N \frac{\mathrm{d}\psi(t)}{\mathrm{d}t} \tag{6.104}$$

This is proportional to the rate of change of occupancy of our tube with time. We can now use this function to describe our shearing behaviour:

$$\sigma(t) = G(t)f_1(\dot{\gamma}t) + \int_0^t m(t)f_1(\dot{\gamma}t)dt \qquad (6.105)$$

This expression can describe the viscosity change with time at a fixed rate and in the limit of long times provides the steady state viscosity at *a* of shear rate. The term $f_1(\gamma)$ is an integral function of the strain. Approximate forms are available, for example:

$$f_1(\gamma) = \frac{5(\gamma^5 - \gamma^4) + a_1\gamma^3 + b_1\gamma}{5(\gamma^6 + c_1\gamma^4 + d_1\gamma^2 + b_1)} \qquad (6.106)$$

with $a_1 = 45.421$, $b_1 = 1269.1$, $c_1 = 10.942$ and $d_1 = 343.57$ and, while awkward in form, this expression contains no mathematical features that are not readily numerically evaluated. This approach has been extended to include normal force and elongational viscosity calculations both with time and the steady state values. The only change required is in the form of the strain function and an additional strain term to determine the normal force. A comparison of the Graessley (G) and Doi–Edwards (DE) models is shown in Figure 6.21 for the viscosity in continuous shear. The reduction in viscosity with shear rate is much more rapid with increasing rate with the DE model than with the G model. The G model has been well tested on near monodisperse molecular weight systems. In addition the DE model has the surprising result of predicting a reduction in stress with increasing rate once a little greater than $\dot{\gamma}\tau_d$. The reason for this is yet to be resolved. It has been observed in pipe flow that with high molecular weight polymers abrupt changes in shear rate occur. It is feasible that the DE model is describing drag reduction behaviour seen in pipe flow or unsteady flow. If we include crudely the role of the Rouse relaxation mechanisms within the tube there is a second process observed, shown as a kink on the DE curve. This is also shown in Figure 6.21 with an arbitrarily selected value of τ_e. The same data is shown as shear stress versus shear rate in Figure 6.22.

It is fairly clear that as τ_e approaches τ_d the role of Rouse relaxation is significant enough to remove the dip altogether in the shear stress–shear rate curve. As the relaxation process broadens, this process is likely to disappear, particularly for polymers with polydisperse molecular weight distributions. The success of the DE model is that it correctly represents trends such as stress overshoot. The result of such a calculation is shown in Figure 6.23.

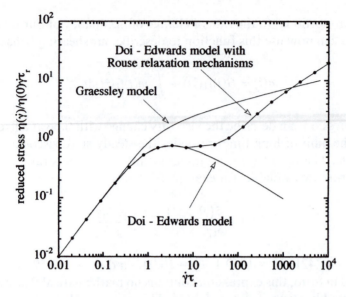

Figure 6.22 *Comparison of Graessley and Doi–Edwards models for normalised stress versus normalised shear rate. Also shown is an estimate of the role of short time Rouse relaxation mechanisms within the tube*

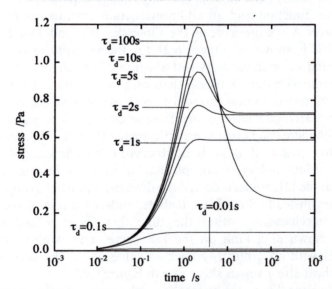

Figure 6.23 *Prediction of stress overshoot for different tube disengagement times. The shear rate used for the calculation was $5\,s^{-1}$*

The qualitative agreement between theory and experiment is very good.[24] The DE approach can be used to provide qualitative information on the effects of branching. In order for a simple unbranched polymer to relax an applied strain we have seen there are two relaxation processes, a

Rouse mechanism within the tube and the disengagement of the polymer from the tube. For a branched polymer the arm is tethered at one end so this restricts motion. In order for disengagement to occur the arm has to retract itself down the tube. This is the dominant timescale and determines the viscosity. We can think of this as akin to an activation energy process giving rise to an exponential dependence in the viscous process. As yet only qualitative agreement has been achieved.

Polydispersity of the molecular weight is not so well described by the DE approach, even qualitatively. In reptation theory the blend of two compatible homopolymers A and B of different molecular weights is given by

$$G^{mix}(t) = x_A G_A(t) + (1 - x_A) G_B(t) \quad \text{where } x_A = \frac{c_A}{c_A + c_B} \tag{6.107}$$

However, experimental observation shows that the relaxation time of the longest process is reduced. Presumably as lower molecular weight species are blended with high molecular weight species the ability for the higher molecular weight species to diffuse becomes important. The constraining tube is weakened and structural reorganisation takes place allowing more rapid relaxation of the stress. A more appropriate shift has been suggested experimentally to be

$$G^{mix}(t) = x_A^2 G_A(t/x_A) + \left(1 + x_A^2\right) G_B(t) \tag{6.108}$$

where polymer A shows the longest relaxation time in isolation from polymer B. The application of reptation dynamics to the non-linear response of semi-dilute and concentrated solution polymers is more problematic. The importance of solution properties on both the hydrodynamics and the tube dimensions is not that well found in the linear regime (Section 5.5.6). These properties will become increasingly important as the strain is increased.

In conclusion, the strength of the reptation approach is not that it provides a rigorous description of the non-linear viscoelasticity of polymers, although agreement can be very good. Its strength is as a way of thinking about polymer rheology problems in terms of the tube and chain dynamics. If we can use our physical intuition to assess how changes in our chemistry affect these we can say something sensible about the accompanying changes in polymer behaviour.

6.5 NOVEL APPLICATIONS

6.5.1 Extension and Complex Flows

The study of extensional flow is becoming one of the most active areas of research in rheology. Although not all extensional flow measurements are non-linear, a large proportion used for chemical engineering studies are. The importance of systems that form fibres, *i.e.* strongly spinnable materials, is of course well recognised. However, there has been a growing recognition that the measurement of fluid systems is an area of enormous practical importance affecting ink jet printers, roll mills, blade coating, curtain coating and many other commercial processes. The development of a microstructural description of the behaviour is hampered by the nature of the experimental data available. This is not a criticism of instrument designers but a reflection of the intrinsic difficulty in making the measurement. The major obstacle is that achieving steady state extensional flow with a wide variety of systems is impractical. The problem is centred on residence time. As an example of this consider a simple couette geometry which we fill with a polymer solution. We can apply a constant shear stress with time and follow the increase in rate until a steady state response is achieved. It may take several hours or days to achieve but nonetheless for many systems a steady response is practically achievable. However, extension presents a more formidable challenge. One way in which an elongational field can be created is to force the fluid through a contracting geometry such as that shown in Figure 6.24.

Suppose we have designed our geometry such that the radius *r* is proportional to the length *l* via a constant *k*:

$$[R(l)]^2 = \frac{k}{l} \tag{6.109}$$

In our thought experiment we can force fluid to flow through this die at a certain volume per unit time or volumetric flow rate *Q* and allow no drag with the walls. This is a total slip condition so no shear will be applied to the fluid. The fluid will travel faster as the geometry contracts so the velocity *V(l)* in a plane across the geometry simply depends upon the area at any length *l*:

$$V(l) = \frac{Q}{\pi[R(l)]^2} = \frac{Ql}{\pi k} \tag{6.110}$$

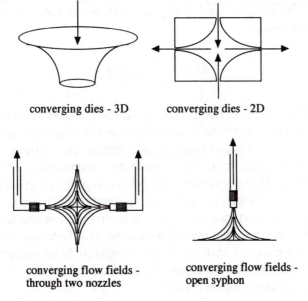

converging dies - 3D

converging dies - 2D

converging flow fields -
through two nozzles

converging flow fields -
open syphon

Figure 6.24 *Flow through contracting geometries and contracting flow fields*

Now the extension rate $\dot{\varepsilon}$ in the geometry is given by the rate of change of velocity with distance:

$$\dot{\varepsilon} = \frac{\mathrm{d}V(l)}{\mathrm{d}l} = \frac{\mathrm{d}}{\mathrm{d}l}\left\{\frac{Ql}{\pi k}\right\} = \frac{Q}{\pi k} \tag{6.111}$$

This expression indicates that for a constant flow rate there is a constant extension rate for this shape of die in a total slip condition. So you might suppose all that is required is to have a lubricated die, a pump and a method of measuring the stress in flow and the extensional viscosity can be obtained from the stress divided by the rate. However, a practical geometry has to have a finite length and the radius has to be large enough to allow our material out of the end! So the fluid can only spend a certain amount of time in the flow – this is called the *residence time*, t_r. This depends on the size of the die and the flow rate. If we divide the total volume of the die V, by the volume flowing per second Q we get the time that the material is resident in the geometry. We can use a volume of rotation integral to determine V between lengths L_1 and L_2:

$$t_r = \frac{V}{Q} = \frac{1}{Q}\pi \int_{L_1}^{L_2} [R(l)]^2 \mathrm{d}l = \frac{\pi k}{Q} \int_{L_1}^{L_2} l^{-1} \mathrm{d}l \tag{6.112}$$

Evaluating the integral and using the extension rate we obtain the following expression:

$$t_r \dot{\varepsilon} = \ln\left(\frac{L_2}{L_1}\right) = w \qquad (6.113)$$

You will notice that w depends only upon the geometry and is invariant with rate. The residence time is the amount of time the material spends at that extension rate; it is our experimental measurement time. It depends on the length of the geometry for a given design constant k. This has the same form as the Hencky strain, where a rod is grabbed at each end and subjected to an exponential increase in length. This gives a uniaxial extension. The importance of this is demonstrated when we consider what would happen to a material measured as a function of extension rate. Suppose we have a material that in extension behaves as a Maxwell model. The extensional viscosity as a function of time and rate as measured in our geometry is given by

$$\eta_e(\dot{\varepsilon}, t) = \eta_e(0)\left[1 - \exp\left(\frac{t_r}{\tau_e}\right)\right] = \eta_e(0)\left[1 - \exp\left(\frac{w}{\dot{\varepsilon}\tau_e}\right)\right] \qquad (6.114)$$

So the Hencky strain determines the apparent viscosity we measure. Each rate defines a different point in time which is not necessarily the steady state value. If $w/\dot{\varepsilon} \gg \tau_e$ the material will appear to be extension thinning and if $w/\dot{\varepsilon} \ll \tau_e$ the viscosity will be constant. If the material displays a significant stress overshoot then it may appear extension thickening. The point we are trying to highlight is that the apparent extensional viscosity is a point on at least a three-dimensional surface where time, rate and viscosity form the axes. It is difficult to design an extensional measurement on viscoelastic systems where this is not a significant effect on the material behaviour. We must also recall that the material in many cases undergoes some pre-shearing in order to reach the extensional testing region. Molecular interpretation is thus difficult. Another factor to be considered is the role of non-extending fields in the flow. The measuring geometry can be designed in a variety of forms; two others are shown as an example in Figure 6.24.

In the opposed jets design fluid is sucked or pumped into a beaker. The profile which develops is dominantly extensional. In the profiled slot design a rectangular channel is designed such that in the total slip condition an extensional flow develops with a constant rate. The pressure is measured at the stagnation point. Other designs include the open syphon, where fluid is sucked from a beaker through a nozzle which is raised above the surface of the liquid. Image analysis of the profile can be used to define the extension rate(s). In each of these flows there is a residence time to consider along with the role of shear and inertia. These complex flow patterns mean that the resulting stress is some combination of three factors:

1. The stress due to the extensional properties of the material at that residence time and range of rates.
2. The stress due to the shear properties of the material at that residence time and range of rates.
3. The kinetic energy term. The fluid accelerates in extensional flow which results in a stress contribution of the order of $\rho_m v^2$ where ρ_m is the density of the material and v its velocity.

These flows are complex combinations of deformations and the need to understand them is important in interpreting behaviour at the molecular level. To illustrate the importance of the strain in elongation, Figure 6.25 shows a plot of data gathered on the opposed jet design. This was for poly(vinyl pyrrolidone) (PVP) with the added surfactant sodium dodecyl sulphate. The surfactant and the polymer give rise to specific interactions, with the surfactant molecules attaching themselves to the chain. The structure probably consists of micelles attached like a string of pearls

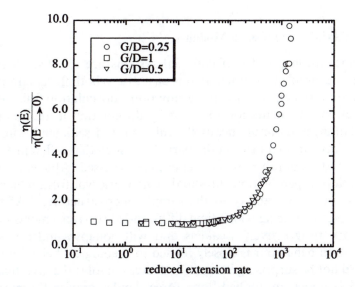

Figure 6.25 *Plot of the normalised extensional viscosity of an 8% PVP solution with 320 mM sodium dodecyl sulphate. The data is plotted versus the reduced extension rate*

along the chain. Eastman[33] performed experiments at a range of extension rates on a range of samples. In the example given below the gap between the nozzles G and their diameter D was varied. The data is only reduced on the x-axis when the extension rate, relaxation time and the strain in the sample are accounted for. These data are plotted as the ratio of the extensional viscosity at a given rate to the extensional viscosity at the low extension rate limit. Molecular interpretation is difficult but clearly the elongational resistance of the chain is significant. It is elongational hardening.

Generalising about the nature of the response of polymeric materials in elongational flow fields is difficult. The coils will tend to align with the flow and begin to stretch. If they are strongly entangled the materials will tend to extension thicken as loops begin to tighten. Transiently cross-linked structures will tend to extension thin as the networks pull apart. However, if the flow of these systems shows a transient elongational stress overshoot this may mask the thinning behaviour observed particularly at short residence times.

In summary this is an area which is changing rapidly in terms of instrumental design and our understanding is developing alongside these changes. For the chemical engineer designing complex flow regimes it is important that it is recognised that the response of a material may not be simply predicted from a shear viscosity.

6.5.2 Uniaxial Compression Modulus

The sedimentation profiles of colloidal systems are affected by many factors. For example in a system that is unstable to both flocculation and sedimentation we can visualise three situations, flocculation followed by sedimentation, sedimentation followed by flocculation in the sediment (consolidation) and simultaneous flocculation and sedimentation. The act of the gravitational field on the particles is such that different Péclet, Reynolds and Deborah number regimes are accessed. Above a critical concentration, a percolation threshold, a network will form which will provide a transient stability to the sample microstructure.[18] After an induction period, during which time diffusion-led rearrangements occur, the network progressively weakens and will collapse under gravity. Sedimentation behaviour is closely related to rheological behaviour and this should not be surprising because both areas involve the investigation of materials under an applied body force. Under gravity the particles sediment and the sedimentation stress is opposed by the excess osmotic pressure in the sediment:

$$\Pi(h) = \frac{mg}{\text{area}} \tag{6.115}$$

where m is the total mass of the particles up to a height h from the bottom of the sediment and the area is that of the tube. The mass of particles is determined from the sum of the masses of all the particles per unit volume $\rho(h)$ multiplied by the volume at height h:

$$\Pi(h) = \Delta\rho g \int_0^h \rho(h)\frac{4}{3}\pi a^3 \frac{V(h)}{\text{area}}\,dh = \Delta\rho g \int_0^h h\varphi(h)\,dh \tag{6.116}$$

Such expressions can be derived from the general treatment by Kynch.[34] The definition of a bulk modulus is

$$K_B = -V\frac{dP}{dV} = -\frac{dP}{d\ln V} \tag{6.117}$$

By analogy we can define a uniaxial modulus of compression $K(\varphi)$ and given that the volume fraction is inversely related to the total volume we get

$$K(\varphi) = \varphi\frac{d\Pi}{d\varphi} = \frac{d\Pi}{d\ln \varphi} \tag{6.118}$$

Now the volume fraction is a function of height and so we can relate our compression modulus to the variation in volume fraction:

$$\int_{\varphi_0}^{\varphi(h)} \frac{K[\varphi(h)]}{\varphi(h)}\,d\varphi(h) = \Delta\rho g \int_0^h h\varphi(h)\,dh \tag{6.119}$$

If the concentration profile can be determined the moduli can be evaluated. In principle there is no reason why this should be a non-linear measurement, it depends upon the magnitude of the gravitational Péclet number. Buscall[35] suggested that a low speed centrifuge could be used to apply different acceleration gradients to the dispersion. If the angular velocity of the rotor is ω_r and if X is the distance from the centre of the rotor to the top of the sediment then the pressure balance equation becomes

$$\Pi(h) = \Delta\rho\omega_r^2 \int_0^h (h + X)\varphi(h)\,dh \tag{6.120}$$

Providing different centrifuge speeds produces a sediment with an equilibrium height h_e and evaluation of the uniaxial modulus of compression is possible. An exact evaluation is somewhat complex but Buscall demonstrated an approximate solution. The mean volume fraction $\bar{\varphi}$ in the sediment is used such that for a starting volume fraction φ_0 with an initial height h_0 the mean value is given by

$$\bar{\varphi} = \varphi_0 \frac{h_0}{h_e} \qquad (6.121)$$

Buscall showed that this results in a simple expression relating the distance X_L from the base of the tube to the centre of the rotor with the equilibrium height for a given angular velocity:

$$K(\bar{\varphi}) \approx \bar{\varphi} \Delta \rho \omega_r h_e^2 \left[\left(X_L - \frac{h_e}{2} \right) \frac{d\omega_r}{dh_e} - \frac{\omega_r}{4} \right] \qquad (6.122)$$

Thus a plot of angular velocity versus h_e for a range of angular velocities can be used to find an approximate value for the uniaxial modulus of compression. The centrifuge is acting as a low stress rheometer. In the limit as $h \to h_0$ then the approximation above becomes an equality so suitable extrapolations of the uniaxial modulus at different heights to zero angular velocity gives the value $K(\bar{\varphi})$. Suppose we assume that our structure is highly ordered, say fcc or bcc in form, with z nearest neighbours spaced a distance r. We could use a cell model for the osmotic pressure as given in Section 5.5. We know that our interparticle spacing is linked by the volume fraction:

$$K(\varphi) = \varphi \frac{d\Pi}{d\varphi} = \varphi \frac{d\Pi}{dr} \frac{dr}{d\varphi} = r \frac{d\Pi}{dr} \qquad \text{where } r = 2a \left(\frac{\varphi_m}{\varphi} \right)^{1/3} \qquad (6.123)$$

So now incorporating a cell model for osmotic pressure:

$$\Pi \approx -\frac{\rho r z}{6} \frac{dV}{dr} \qquad (6.124)$$

Substituting into the expression for the uniaxial modulus and retaining only the most significant terms we get

$$K(\varphi) \approx r \frac{d}{dr} \left\{ -\frac{\rho r z}{6} \frac{dV}{dr} \right\} \approx \frac{\rho r^2 z}{6} \frac{d^2 V}{dr^2} \qquad (6.125)$$

Our expression for the high frequency shear modulus is now (Section 5.5):

$$G(\infty) = \frac{3\varphi_m z}{32r} \frac{d^2 V}{dr^2} \qquad (6.126)$$

Thus we would expect $K(\varphi)/G(\infty) \approx 3\text{--}4$. Experimentally Buscall has shown close agreement between these two properties. This again acts to confirm the rheological nature of this study. As the stress is increased the applied field can cause significant changes to the structure. Under these circumstances the experiment has entered a non-linear regime and these simple relationships will begin to fail.

6.5.3 Deformable Particles

Deformable particles cover a wide range of materials from human erythrocytes to emulsions and microgels. Rheological measurements on these systems show a high sensitivity to the chemical environment. Energy dissipation mechanisms will include the hydrodynamic drag of the particles through the medium and interparticle forces. In addition they will include mechanisms associated with deforming the shape of the particles and internal circulation within them where possible. Thus the effect of the applied field is a subtle balance between the internal and external structural rearrangements that are possible.

The most widely studied deformable systems are emulsions. These can come in many forms, with oil in water (O/W) and water in oil (W/O) the most commonly encountered. However, there are multiple emulsions where oil or water droplets become trapped inside another drop such that they are W/O/W or O/W/O. Silicone oils can become incompatible at certain molecular weights and with different chemical substitutions and this can lead to oil in oil emulsions O/O. At high concentrations, typical of some pharmaceutical creams, cosmetics and foodstuffs the droplets are in contact and deform. Volume fractions in excess of 0.90 can be achieved. The drops are separated by thin surfactant films. Self-bodied systems are multicomponent systems in which the dispersion is a mixture of droplets and precipitated organic species such as a long chain alcohol. The solids can form part of the stabilising layer – these are called Pickering emulsions.

The surface characteristics of these species are determined by the particulates and stress transfer across the membrane will tend to be low, reducing internal circulation within the drop. The structure of the interface surrounding the drop plays a significant role in determining the characteristics of the droplet behaviour. We can begin our consideration of emulsion systems by looking at the role of this layer in determining linear viscoelastic properties. This was undertaken by

Princen and Kiss[36] and also by Mason *et al.*[37] They suggested a simple semi-empirical linearisation of $G(0)$ the static modulus with volume fraction:

Princen

$$G(0) = \frac{A\gamma_{12}\varphi^{1/3}}{a_{32}}(\varphi - \varphi_m) \qquad (6.127)$$

Mason

$$G(0) = \frac{A\gamma_{12}\varphi}{a_{32}}(\varphi - \varphi_m) \qquad (6.128)$$

where A is a constant between typically between 1 and 2. The radius a_{32} is the third moment to second moment average and γ_{12} is the interfacial tension between the droplet and the continuous phase. The term φ_m is the maximum packing fraction before the drops distort. The basic idea underlying this model is that at a critical concentration the droplets pack together and thin films begin to form between them. As the concentration is increased further the droplets deform. This type of emulsion is called a *high internal phase emulsion* (HIPE) or sometimes a *gel emulsion*. If a strain is applied this results in droplet distortion and an increase in the film area which is opposed by the interfacial tension. Taylor and Pons[38] have demonstrated the excellence of these expressions at linearising the data for G', the storage modulus, for a wide range of systems. They have also demonstrated that systematic blending of stabilisers can be used to tune the elasticity of an emulsion in keeping with the spirit behind the approach. The surprising feature of the Princen and Mason expressions is that despite recorded differences in the polydispersity of droplet size the equations work very well. This factor is not well accounted for in the model. This suggests that some other common feature has a strong influence on the elasticity.

One factor which could be fairly constant despite size polydispersity is the monodispersity of the film dimensions. Taylor and Pons have tentatively suggested that this has an effect on the constant A where molecular weight polydispersity and impurities in the stabilisers apparently promote an increase in its value. There has only been a little work on modelling emulsion systems in comparison to that on their rigid counterparts. Many HIPEs and stable emulsions are formed from systems where strong association occurs in the adsorbing layer. In some cases a liquid crystalline layer forms around the surface of the drop, as has been suggested by Friberg.[39] Such a system can be produced using an ABA block copolymer of poly(12-hydroxystearic acid) (PHS) (A) and

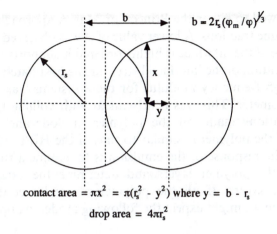

contact area $= \pi x^2 = \pi(r_s^2 - y^2)$ where $y = b - r_s$

drop area $= 4\pi r_s^2$

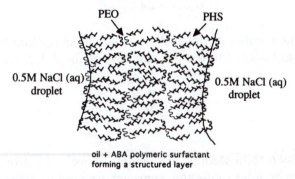

oil + ABA polymeric surfactant
forming a structured layer

Figure 6.26 *Structure of the layer surrounding an emulsion drop and a schematic of the lens overlap area*

poly(ethylene oxide) (PEO) (B). These produce stable HIPEs formed from 0.5M sodium chloride solution with dodecane as the continuous phase.[40] These are strongly viscoelastic above a packing fraction of $\varphi_m = 0.64$. The structure of the associating layer is shown in Figure 6.26.

These systems are viscoelastic, showing a simple Maxwell process added onto a long broad relaxation process with a lower magnitude. There is strong evidence to suggest that the Maxwell process is due to the film layer. The concentration of the polymer at the interface and in between the drops was determined. This polymer concentration was studied in dodecane without the presence of the aqueous drops and it was found to be birefringent, suggestive of a neat phase. It also showed Maxwell behaviour at very similar relaxation times to that seen for the emulsion system. This suggests that it is the film elasticity that is responsible for the Maxwell relaxation processes. Stress relaxation studies on the emulsion indicate that there is an apparent $G(0)$ value in addition to a high frequency modulus. This static modulus conforms reasonably well to either the Princen or Mason approach except at the highest volume fractions. A lower value of A was observed and given that the structure at the interface is highly ordered it supports the notion that A is an indication of the interfacial structure as postulated by Taylor and Pons. A high frequency modulus for these systems was found using a pulse shearometer. This was nearly linear with volume fraction. From the observations made on the polymer in dodecane we know the elasticity of the polymer is similar to that of the HIPE. One method of modelling the response of the emulsion is to assume a rubber elasticity model for the polymer layers and determine the extent of contact between the layers. If all the polymer molecules contribute to the elasticity then we might expect the following model to apply:

$$G(\infty) = 2A_c \rho_e k_B T \qquad (6.129)$$

where ρ_e is the number of elastically effective links per unit volume and A_c is a constant related to the functionality of the link f_L as defined by Graessley:[41]

$$A_c = \frac{f_L - 2}{f_L} \qquad (6.130)$$

However, only certain sections of the films between the drops will be in contact. Thus in order to use this approach we need to define the region of overlap between the layers on the drops, the number of regions which overlap and the number of chains in contact. We can approximate the area of the overlap of the layers on neighbouring drops by assuming that as the layers interpenetrate a lens-like area is formed (see Figure 6.26 for example). The area of this relative to the drop area is

$$C_a = \frac{\text{area of lens}}{\text{area of drop}} = \frac{1}{4}\left\{1 - \left[2\left(\frac{\varphi_m}{\varphi_e}\right)^{1/3} - 1\right]^2\right\} \qquad (6.131)$$

where φ_e is the volume fraction of the drop and the stabilising shell. If n_m is the number of molecules per droplet and every molecule is elastically effective, a reasonable assumption with liquid crystalline order, then the number of effective chains per unit volume is simply calculated. It is given by the number of interactions per unit volume multiplied by n_m:

$$\rho_e = n_m C_a \left(\frac{3\varphi_c}{4\pi a_c^3}\right) \qquad (6.132)$$

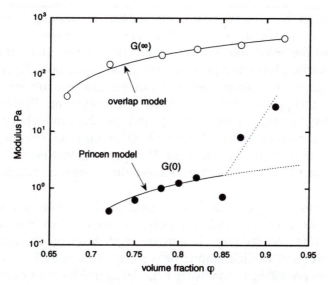

Figure 6.27 *Plot of the high frequency shear modulus G(∞) and the static modulus G(0) versus volume fraction. The points are the experimental data, the solid lines represent the models*

Here φ_c is the volume fraction of the core and a_c its radius. This equation has not been widely tested owing to a paucity of data. Thorough characterisation allows all the terms to be determined except φ_m and f_L. The packing fraction can be found by extrapolation to zero concentration of a plot of the high frequency shear modulus as a function of volume fraction since this corresponds to the volume fraction before the chains come into contact. The functionality of the link can be used as an adjustable parameter. For the system here a good fit is found with $f_L = 8/3$, as shown in Figure 6.27.

A universal model for the viscoelastic response of HIPEs is unlikely to be satisfactory without the inclusion of the interfacial film effects. For example Solans *et al.*[42] suggested for systems stabilised by non-ionic surfactants, the Princen equation for $G(0)$ can be used to describe the high frequency shear modulus rather than the static modulus. The difference between their systems and a polymer stabilised system discussed above is presumably due to the structure of the film between the drops and its viscoelastic characteristics. Changing the temperature can cause changes in elasticity, relaxation times and interfacial film properties. It can also lead to phase inversion with a W/O emulsion becoming O/W. This change leads to marked differences in viscosity, thus measuring the viscosity at a fixed shear rate as the temperature is changed can be used to indicate the phase inversion temperature. As the concentration of the emulsion is reduced below the maximum packing fraction the elasticity is lost and the material becomes pseudoplastic. For water in oil systems the behaviour observed is dictated by the extent of aggregation, the surface layer and polydispersity. Large size polydispersity seen in many emulsion systems tends to cause an emulsion to shear thin less rapidly with applied stress than hard sphere systems. Polyelectrolyte stabilised systems can show both high and low shear rate plateaus which are less commonly detected with W/O emulsions. The high shear viscosity can conform to a Krieger–Dougherty model normally with a slightly lower intrinsic viscosity to allow for internal circulation within the drop.

At present our understanding of emulsion behaviour is not as well developed as that of particulate or polymer systems. Part of the difficulty in correlating the rheology lies in the high level of characterisation required in order to differentiate between systems as well as the greater difficulty in preparing monodisperse model emulsions than rigid particulate systems. However, this is understandable because emulsion characterisation can be formidable.

6.6 REFERENCES

1. N. Casson, in *Rheology of Disperse Systems*, ed. C.C. Mill, Pergamon Press, New York, 1959, p. 84.
2. M.M. Cross, *J. Coll. Interface Sci.* 1965, **20**, 417.
3. M.E. Woods and I.M. Krieger, *J. Coll. Interface Sci.* 1970, **34**, 91.
   Y.S. Papir and I.M. Krieger, *J. Coll. Interface Sci.* 1970, **34**, 126.
4. E.C. Bingham, *Fluidity and Plasticity*, McGraw-Hill, New York, 1922.
5. W.P. Cox, and E.H. Merz, *J. Polym. Sci.* 1958, **28**, 619.
6. M. Doi and S.F. Edwards, *The Theory of Polymer Dynamics*, Oxford University Press, Oxford, 1986.
7. I.M. Krieger and T.J. Dougherty, *Trans. Soc. Rheol.* 1959, **3**, 137.
8. R. Buscall, J.W. Goodwin, R.H. Ottewill and Th.F. Tadros, *J. Coll. Interface Sci.* 1982, **85**, 78.
9. A.J. Bradbury, J.W. Goodwin, and R.W. Hughes, *Langmuir*, 1992, **8**, 2863.
10. R.B. Bird, R.C. Armstrong and O. Hassanger, *Dynamics of Polymeric Liquids*, Vol. 1, Wiley, New York, 1977.
11. N.A. Frankel and A. Acrivos, *Chem. Eng. Sci.* 1967, **22**, 847.
    J.D. Goddard, *J. Non-Newtonian Fluid Mech.* 1977, **2**, 169.
12. J.R. Melrose, J.H. van Vliet, L.E. Silbert, R.C. Ball and R. Farr, in *Modern Aspects of Colloidal Dispersions*, eds R.H. Ottewill and A.R. Rennie, Kluwer Publications, Dordrecht, 1998, p. 113.
13. A.S. Michaels and J.C. Bolger, *Ind. Eng. Chem., Fundam.* 1962, **1**, 153.
14. J.W. Goodwin, R.W. Hughes, H.M. Kwaambwa and P.A. Reynolds, *Coll. & Surf.* 2000, **161**, 361.
15. R.J. Hunter, *Foundations of Colloid Science*, Vols 1 and 2, Oxford University Press, Oxford, 1995.
16. W.B. Russel, D.A. Saville and W.R. Scholwater, *Colloidal Dispersions*, Cambridge University Press, Cambridge, 1991.

17. P.D.A. Mills, J.W. Goodwin and B. Grover, *Colloid Polym. Sci.* 1991, **269**, 949.
18. J.W. Goodwin and R.W. Hughes, *Adv. Coll. Interface Sci.* 1992, **42**, 303.
19. J.W. Goodwin, Ph.D thesis, University of Bristol, 1972.
20. E. Dickinson, *An Introduction to Food Colloids*, Oxford University Press, Oxford, 1992.
21. H.M. Lindsay and P.M. Chaikin, *J. Phys. (Paris) C3*, 1985, **46**, 269.
22. R. Buscall, in *Colloidal Polymer Particles*, eds J.W. Goodwin and R. Buscall, Academic Press, London, 1995, p. 49.
23. C.F. Zukoski and L. Chen, *J. Chem. Soc., Faraday Trans.* 1990, **86**, 2629.
24. M. Doi and S.F. Edwards, *J. Chem. Soc., Faraday Trans. 2*, 1978, **74**, 1818; 1978, **75**, 38; and reference [6].
25. P.J. Carreau, Ph.D. thesis, University of Wisconsin, 1968.
26. G. Chauveteau, *J. Rheol.* 1982, **26**, 111.
27. S.A. Keeping, Ph.D thesis, University of Bristol, 1989.
28. W.W. Graessley, *J. Chem. Phys.* 1967, **47**, 1942.
29. J.D. Ferry, *Viscoelastic Properties of Polymers*, 3rd edn, Wiley, New York, 1980.
30. W.W. Graessley, *J. Chem. Phys.* 1971, **54**, 5143.
31. M.C. Williams, *AIChE J.* 1966, **12**, 1064.
32. W.W. Graessley, *Adv. Polym. Sci.* 1974, **16**, 1.
33. J. Eastman. Ph.D thesis, University of Bristol, 1995.
34. G.J. Kynch, *Trans. Far. Soc.* 1952, **48**, 166.
35. R. Buscall, *Colloids Surf.* 1982, **5**, 269.
36. H.M. Princen and A.D. Kiss, *J. Coll. Interface Sci.* 1986, **112**, 427.
37. T.G. Mason, J. Bibette and D.A. Weitz, *Phys. Rev. Lett.* 1995, **75**, 2051.
38. P. Taylor and R. Pons, in *Modern Aspects of Colloidal Dispersions*, eds R.H. Ottewill and A.R. Rennie, Kluwer Publications, 1998 p. 225.
39. S. Friberg, L. Mandell and M. Larson, *J. Coll. Interface Sci.* 1969, **29**, 155.
    S. Friberg, *J. Coll. Interface Sci.* 1977, **37**, 291.
    S. Friberg, P.O. Jansson, E. Cederberg, *J. Coll. Interface Sci.* 1976, **55**, 614.
40. R.W. Hughes, Ph.D thesis, University of Bristol, 1988.
41. W.W. Graessley, *Macromolecules*, 1975, **8**, 186; 1975, **8**, 865.
42. C. Solans, R. Pons and H. Kunieda, in *Modern Aspects of Emulsion Science*, ed. B. Binks, Royal Society of Chemistry, Cambridge, 1998. Chapter 11.

Subject Index